TIDES AND TIDAL STREAMS

TIDES
AND
TIDAL STREAMS

A MANUAL COMPILED FOR THE
USE OF SEAMEN

BY

COMMANDER H. D. WARBURG, R.N.
(RETIRED)

SUPERINTENDENT OF TIDAL WORK, HYDROGRAPHIC DEPARTMENT, ADMIRALTY

CAMBRIDGE

AT THE UNIVERSITY PRESS

1922

CAMBRIDGE
UNIVERSITY PRESS

University Printing House, Cambridge CB2 8BS, United Kingdom

Cambridge University Press is part of the University of Cambridge.

It furthers the University's mission by disseminating knowledge in the pursuit of education, learning and research at the highest international levels of excellence.

www.cambridge.org
Information on this title: www.cambridge.org/9781107559936

© Cambridge University Press 1922

First published 1922
First paperback edition 2015

A catalogue record for this publication is available from the British Library

ISBN 978-1-107-55993-6 Paperback

PREFACE

Many books on the tides have been published; the majority are either of a highly technical and scientific nature or are written with the object of proving or disproving a particular theory; the primary object of the present Manual is to explain the tides to seamen.

The Manual is entirely unofficial; in the writer's opinion tidal instruction as given in the Navigation Schools, both of the Royal Navy and of the Mercantile Marine, is incomplete and faulty and has undergone less improvement than any other branch of Navigation; the modern seaman, who learns that tides are due to the moon's attraction and who finds the time of high water by means of the time of the moon's transit and the Establishment of the port, is, in fact, in all essentials, receiving the teaching and using the methods of his ancestors of the seventeenth century. Where the tide is entirely semi-diurnal these methods have the excuse that a result correct to within an hour or so is obtained, but where the diurnal waves are large results are so wildly inaccurate that the time found may be more nearly that of low water than of high water.

In this Manual, therefore, an attempt has been made to give an adequate explanation of the causes of all tidal phenomena, and, whilst explaining the correct methods of using the tidal information now available, to introduce improved methods of giving such information by means of harmonic constants, and to explain how, from these constants, the height of the tide may be calculated, at any port in the world and at any moment, with the necessary degree of accuracy. It is hoped that this method will meet with both official and non-official approval and that, in course of time, harmonic constants will be given in all tide tables instead of, or in addition to, the non-harmonic constants now given.

Each chapter in the Manual is, as far as possible, complete in itself and, in order to simplify reference to other portions where required, is sub-divided into numbered sections; no index is given, for each chapter must be regarded as a whole and an incorrect impression of tidal phenomena might be obtained from reading one or two sections only.

PREFACE

The explanation of the cause of tides, given in Chapter I, is adapted from the *Manual of Tides* by Rollin A. Harris (Government Printing Office, Washington) by permission of the Director of the United States Coast and Geodetic Survey; the explanation of harmonic methods in Chapter VII is adapted from *Winds, Weather, Currents, Tides and Tidal Streams in the East Indian Archipelago* by J. P. Van der Stok, Ph.D., Director of the Royal Netherlands Meteorological and Magnetic Observatory (Government Printing Office, Batavia); my thanks are especially due to Dr Van der Stok for permission to make use of the published results of his tidal work, without which it would have been difficult to introduce new and improved methods to British seamen. The tide curves (Figures 5 and 6) were supplied by Messrs Ed. Roberts & Son, the latter by permission of the Superintendent of the Canadian Tidal and Current Survey.

H. D. WARBURG.

January 1922.

CONTENTS

*This map is available for download from www.cambridge.org/9781107559936

CHAPTER I

The Tide-raising Forces. Lunar and Solar Tides.
The Combined Wave

§ 1. Periodic vertical movements of the waters on the earth's surface are called *Tides*, periodic horizontal movements are called *Tidal Streams*; this distinction is of importance and tidal streams must never be referred to as tides.

Tidal movements are essentially periodic; waves due to earthquakes or submarine volcanic eruptions, popularly but erroneously called tidal waves, are therefore not tides, and oceanic currents, flowing always in one direction, are not tidal streams.

The tide rises to a maximum height, called *High water*, and falls to a minimum height, called *Low water*; the difference between the heights of high water and low water is the *Range* of the tide.

Mean sea level at any place is the average water level at that place.

§ 2. All heavenly bodies attract each other, the force of attraction varying directly with the mass of the attracting body and inversely as the square of its distance.

Though the earth is attracted by all heavenly bodies, the attraction of the planets and fixed stars is so small as to have no effect in raising tides, and the moon and sun only need be considered.

§ 3. The tides as they occur are greatly affected by variations in the depth of the sea, by shallow water, by land and by frictional resistance; in examining the forces to which the tides are due, and the tides to be expected, the earth will therefore be considered to consist of a solid spherical nucleus entirely surrounded by fluid of great depth and devoid of frictional resistance—the perfect tidal earth—thus reducing the subject to be examined to the astronomical tide only, unaffected by terrestrial conditions. Further the earth, moon and sun will first be considered as fixed in space, the moon and the sun being situated in the plane of the earth's equator. Under these conditions the tide becomes a deformation only in the shape of the surrounding fluid, for it is evident that, without movement, there can be no visible tide.

§ 4. The moon attracts every particle of the earth, solid or fluid, the force acting on each particle is directed towards the moon's centre, and is inversely proportional to the square of the distance between the particle and the moon's centre.

The average attraction of the moon on all the solid and liquid particles of which the earth is formed is exerted on a particle at the average position of all the particles, i.e. on a particle at the earth's centre; the difference between the attraction on this particle and on each other particle is the *Tide-raising force* of the moon.

§ 5. In Figure 1, let M be the moon's centre; E a particle at the earth's centre; $NOSW$ the surface of the fluid exterior of the earth, O and W being particles on the line joining ME or on that line produced, N and S particles at the poles of a great circle perpendicular to the line joining ME.

If the line OM represent, both in direction and magnitude, the force of the moon's attraction on O, then

$$\text{Force of moon's attraction on } E = OM \times \frac{OM^2}{EM^2} = OM \left(1 + \frac{OE}{OM}\right)^{-2}$$

$$= OM - 2OE, \text{ nearly,}$$

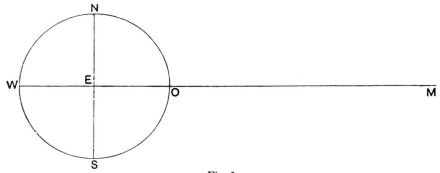

Fig. 1

and

$$\text{Moon's tide-raising force at } O = OM - (OM - 2OE)$$

$$= + 2OE, \text{ acting towards the moon.} \quad (1)$$

Similarly, if WM represent, both in direction and magnitude, the force of the moon's attraction on W, then

$$\text{Force of moon's attraction on } E = WM \times \frac{WM^2}{EM^2} = WM + 2WE, \text{ nearly,}$$

and

$$\text{Moon's tide-raising force at } W = WM - (WM + 2WE)$$

$$= - 2WE, \text{ acting away from the moon.} \quad (2)$$

But OE and WE are equal, and there is thus a tide-raising force acting towards the moon on the side of the earth facing the moon, and a nearly equal force acting away from the moon on the side of the earth turned away from the moon.

§ 6. In Figure 2, let *NOSW* be particles on the surface of the fluid exterior of the earth, *E* a particle at the earth's centre and *M* the moon's centre, as in Figure 1, and let *P* be any other particle on the surface of the fluid exterior of the earth. Pet *PV* be perpendicular to *EM* and *VK* = 2*EV*.

If the line *PM* represent, both in direction and magnitude, the force of the moon's attraction on *P*, then

$$\text{Force of moon's attraction on } E = PM \times \frac{PM^2}{EM^2},$$

but *VM* is nearly equal to *PM* and *EM* = *EV* + *VM*.

Therefore:

$$\text{Force of moon's attraction on } E = VM \times \frac{VM^2}{(EV + VM)^2}$$

$$= VM - 2EV, \text{ nearly,}$$

$$= KM, \text{ by construction.}$$

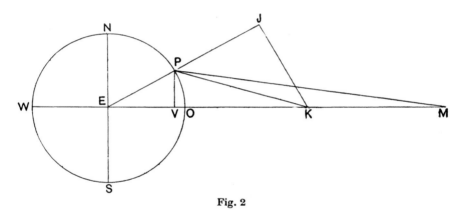

Fig. 2

KM therefore represents, both in direction and magnitude, the force of the moon's attraction on *E*, and *PK*, completing the triangle of forces, represents, both in direction and magnitude, the moon's tide-raising force at *P*.

§ 7. If *EP* be produced to *J*, so that *KJ* is perpendicular to *EJ*, *PJ* represents, both in direction and magnitude, the vertical component of the moon's tide-raising force at *P*. Displacement of fluid at different points on the earth's surface is proportionate to the total tide-raising force, but the tide (see § 1) is vertical displacement only and is therefore proportionate to the vertical component, and the tide produced by the moon's tide-raising force for various positions of *P* varies as the length of *PJ*.

§ 8. If *PM* = *D* and *EM* = *d*, if μ be the force of attraction between particles at unit distance and if *M* be the moon's mass, then the force of the moon's attraction on P, represented by the line PM, is $\dfrac{\mu M}{D^2}$ and the force of the moon's attraction on E, represented by the line KM, is $\dfrac{\mu M}{d^2}$.

On the same scale, and because, owing to the great distance of the moon, PM and VM are nearly equal, VK represents the difference between the force of the moon's attraction on P and on E, or $\left(\dfrac{\mu M}{D^2} - \dfrac{\mu M}{d^2}\right)$.

If the earth's radius $= r$, if the angle $PEV = \alpha$, and the angle $PKV = \beta$, then, by construction, $\tan \beta = \frac{1}{2} \tan \alpha$, the angle $JPK = (\alpha + \beta)$, and

$$\text{Force represented by } PK \text{ is } \left(\frac{\mu M}{D^2} - \frac{\mu M}{d^2}\right)\frac{1}{\cos \beta},$$

$$\text{Force represented by } PJ \text{ is } \left(\frac{\mu M}{D^2} - \frac{\mu M}{d^2}\right)\frac{\cos (\alpha + \beta)}{\cos \beta}$$

$$= \mu M \left(\frac{1}{D^2} - \frac{1}{d^2}\right)\frac{\cos \alpha \cos \beta - \sin \alpha \sin \beta}{\cos \beta}$$

$$= \mu M \left(\frac{1}{d^2} + \frac{2r \cos \alpha}{d^3} - \frac{1}{d^2}\right)(\cos \alpha - \sin \alpha \tfrac{1}{2} \tan \alpha)$$

$$= \frac{\mu M 2r \cos \alpha}{d^3} \cdot \frac{3 \cos^2 \alpha - 1}{2 \cos \alpha}$$

$$= \frac{\mu M r}{d^3}(3 \cos^2 \alpha - 1). \tag{3}$$

By the laws of gravitation, where E is the earth's mass and g the force of gravitation, $\mu = \dfrac{r^2 g}{E}$.

Therefore PJ varies as $\dfrac{M}{E}\left(\dfrac{r}{d}\right)^3 g\,(3 \cos^2 \alpha - 1)$ $\tag{4}$

and (a) The change in level of the surface of the fluid at any place on the earth's surface varies directly as the mass of the disturbing body and inversely as the cube of its distance.

(b) The change in level of the surface of the fluid at any place on the earth's surface varies as $(3 \cos^2 \alpha - 1)$, where α is the angle at the centre of the earth between the centre of the disturbing body and the place.

These laws are general, and apply both to the solar and to the lunar tides.

§ 9. The earth, moon and sun have hitherto been considered as fixed in space, the change in surface level of the fluid is therefore a change in shape of the fluid exterior of the earth only.

If now the earth be supposed to rotate on its axis whilst the tide-raising body remains fixed, it is evident that, whereas the deformation in the shape of the fluid will remain fixed with reference to the body, a place on the earth's surface will change its position with reference to the body, and a tide will occur with an interval equal to one half the period of rotation of the earth between successive high waters.

§ 10. The movements of the earth, other than rotation, and of the moon must now be considered.

The earth moves round the sun in an elliptical path inclined at an angle of about $23\frac{1}{2}°$ to the plane of its equator; the eccentricity of the earth's orbit is about 0·016. The distance of the earth from the sun is measured by the sun's horizontal parallax, and the angular distance of the sun from the plane of the earth's equator by the sun's declination.

The moon moves round the earth in an elliptical path, inclined, on the average, at an angle of about $23\frac{1}{2}°$ to the plane of the earth's equator; the eccentricity of the moon's orbit is about 0·055. The distance of the moon from the earth is measured by the moon's horizontal parallax, and the angular distance of the moon from the plane of the earth's equator by the moon's declination.

The angular distance of the sun from the meridian of the place is the sun's hour angle, the angular distance of the moon from the meridian of the place is the moon's hour angle.

Declination is measured at the centre of the earth, hour angle at the poles.

Horizontal parallax is the angle at the centre of the moon or sun subtended by the semi-diameter of the earth.

§ 11. The tide, either lunar or solar, varies as $(3 \cos^2 \alpha - 1)$ and may therefore be said to equal $t (3 \cos^2 \alpha - 1)$.

By spherical trigonometry, where l is the latitude of the place, f the declination of the body and h its hour angle,

$$\cos \alpha = \cos l \cos f \cos h \pm \sin l \sin f,$$

the sign being plus when latitude and declination are both North or both South, minus in other cases.

Therefore:

$$t (3 \cos^2 \alpha - 1) = t \{3 (\cos l \cos f \cos h \pm \sin l \sin f)^2 - 1\}$$
$$= t (3 \cos^2 l \cos^2 f \cos^2 h \pm 6 \sin l \cos l \sin f \cos f \cos h$$
$$+ 3 \sin^2 l \sin^2 f - 1)$$
$$= \tfrac{3}{2}t \cos^2 l \cos^2 f \cos 2h \pm 3t \sin l \cos l \sin 2f \cos h$$
$$+ 3t \sin^2 l \sin^2 f - t + \tfrac{3}{2}t \cos^2 l \cos^2 f$$
$$= \tfrac{3}{2}t \cos^2 l \cos^2 f \cos 2h \pm 3t \sin l \cos l \sin 2f \cos h$$
$$+ \tfrac{1}{2}t (3 \sin^2 l - 1) (3 \sin^2 f - 1), \qquad (5)$$

and the tide, either lunar or solar, at any place on the earth's surface, thus consists of three portions:

(*a*) A semi-diurnal portion, varying as the cosine squared of the declination multiplied by the cosine of twice the hour angle. This portion is semi-diurnal, for cos 2h is + when h is between 315° and 45° and between 135° and 225°, − at other times, high water therefore occurs twice a day, when hour angle is 0° and 180°, and low water twice a day, when hour angle is 90° and 270°.

(*b*) A diurnal portion, varying as the sine of twice the declination multiplied by the cosine of the hour angle. This portion is diurnal, for cos h is + when h is between 270° and 90° and − when h is between 90° and 270°, high water therefore occurs once a day, when hour angle is 0°, and low water once a day, when hour angle is 180°.

(*c*) A portion varying only with the declination. $(3 \sin^2 f - 1)$ is + when declination is greater than about 11°, − when declination is less than that quantity, the whole tide wave is therefore raised, and mean sea level is high, when declination is great and is correspondingly lowered when declination is small.

§ 12. Values for t of the solar and lunar tides may be found as follows: When the moon or sun is on the meridian of the place and $\alpha = 0$

$$t \,(3 \cos^2 \alpha - 1) = 2t = \frac{M}{E} \left(\frac{r}{d}\right)^3 r$$

$$= \frac{1}{80} \times \frac{1}{(60)^3} \times 4000 \times 5280$$

$$= 1\cdot 22 \text{ ft., for the moon.} \tag{6}$$

Therefore t of the lunar tide = 0·61 ft.

$$t \,(3 \cos^2 \alpha - 1) = 2t = \frac{S}{E} \left(\frac{r}{d}\right)^3 r$$

$$= 330{,}000 \times \frac{1}{(23{,}000)^3} \times 4000 \times 5280$$

$$= 0\cdot 56 \text{ ft., for the sun.} \tag{7}$$

Therefore t of the solar tide = 0·28 ft.

§ 13. The eccentricity of the moon's orbit is 0·055, and t varies as $\left(\dfrac{r}{d}\right)^3$. Therefore

lunar tide at perigee: t of the lunar tide $= \dfrac{1}{(0\cdot 945)^3} : 1 = 1\cdot 18 : 1$ (8)

and

lunar tide at apogee: t of the lunar tide $= \dfrac{1}{(1\cdot 055)^3} : 1 = 0\cdot 85 : 1.$ (9)

The eccentricity of the earth's orbit is 0·016, and t varies as $\left(\dfrac{r}{d}\right)^3$.

Therefore

solar tide at perihelion: t of the solar tide $= \dfrac{1}{(0·984)^3} : 1 = 1·05 : 1$ (10)

and

solar tide at aphelion: t of the solar tide $= \dfrac{1}{(1·016)^3} : 1 = 0·95 : 1.$ (11)

From expression (4) the range of the tide varies as $\left(\dfrac{r}{d}\right)^3$ and from § 11 (a) the range of the semi-diurnal portion varies as $\cos^2 f$; the diurnal portion, § 11 (b), and the portion which varies only with the declination, § 11 (c), do not affect the mean tidal movement on any day (see § 16), therefore tidal range varies as $\left(\dfrac{r}{d}\right)^3 \cos^2 f$.

$\dfrac{r}{d} = \tan$ horizontal parallax; the moon's mean horizontal parallax is about $57'$ and declination about $14°5$, the sun's mean horizontal parallax is $8''8$ and declination about $14°5$, therefore on any day:

Range of lunar tide $=$ mean range $\times \dfrac{\tan^3 H.P.\ \cos^2 F}{\tan^3 57'\ \cos^2 14°5}$,   (12)

Range of solar tide $=$ mean range $\times \dfrac{\tan^3 h.p.\ \cos^2 f}{\tan^3 8''8\ \cos^2 14°5}$,   (13)

where $H.P.$ is the moon's and $h.p.$ the sun's horizontal parallax on the day, F the moon's and f the sun's declination on the day.

§ 14. Assuming horizontal parallax to be at its mean value, the height of the tide above or below the undisturbed surface of the fluid exterior of the earth is given by expression (5).

From this expression, the heights of high water (H.W.) and low water (L.W.), where $h = 0°$, $90°$, $180°$ and $270°$, are:

H.W. $= \frac{3}{2}t \cos^2 l \cos^2 f + 3t \sin l \cos l \cos 2f + \frac{1}{4}t (3 \sin^2 l - 1)(3 \sin^2 f - 1).$ (14)

L.W. $= -\frac{3}{2}t \cos^2 l \cos^2 f \qquad\qquad + \frac{1}{4}t (3 \sin^2 l - 1)(3 \sin^2 f - 1).$ (15)

H.W. $= \frac{3}{2}t \cos^2 l \cos^2 f - 3t \sin l \cos l \cos 2f + \frac{1}{4}t (3 \sin^2 l - 1)(3 \sin^2 f - 1).$ (16)

L.W. $= -\frac{3}{2}t \cos^2 l \cos^2 f \qquad\qquad + \frac{1}{4}t (3 \sin^2 l - 1)(3 \sin^2 f - 1).$ (17)

From these expressions:

(a) There will be two high waters of unequal height each day, and

Mean high water $= \frac{3}{2}t \cos^2 l \cos^2 f + \frac{1}{4}t (3 \sin^2 l - 1)(3 \sin^2 f - 1).$ (18)

(b) There will be two low waters of equal height each day, and

Mean low water $= -\frac{3}{2}t \cos^2 l \cos^2 f + \frac{1}{4}t (3 \sin^2 l - 1)(3 \sin^2 f - 1).$ (19)

(c) Height of mean sea level $= \frac{1}{4}t (3 \sin^2 l - 1)(3 \sin^2 f - 1).$ (20)

Expressions (18) to (20) give heights above or below the undisturbed surface of the fluid exterior of the earth.

Note. From expressions (14) to (17), if

$$3t \sin l \cos l \sin 2f > \tfrac{3}{2}t \cos^2 l \cos^2 f, \quad \text{or if} \quad \tan l > \tfrac{1}{4} \cot f$$

the high water given by expression (16) will be below mean sea level; and if

$$3t \sin l \cos l \sin 2f > 3t \cos^2 l \cos^2 f, \quad \text{or if} \quad \tan l > \tfrac{1}{2} \cot f$$

the high water given by expression (16) will be below the low waters given by expressions (15) and (17), that is to say there will be only one high water, expression (14), and one low water, expression (16), in each day, and *Single Day* tides will occur.

As the moon's maximum declination is about $28\tfrac{1}{2}°$ and the sun's about $23\tfrac{1}{2}°$, single day lunar tides may occur in latitudes greater than about $43°$ and single day solar tides in latitudes greater than about $49°$.

§ 15. From expression (20) mean sea level varies with both latitude and declination.

In latitude $0°$, when the body is in its mean position and declination is $0°$

> Mean sea level $= t$, above the undisturbed surface of the fluid. (21)

In latitude $90°$, when declination is $0°$

> Mean sea level $= -\tfrac{1}{2}t$, below the undisturbed surface. (22)

Tidal influence therefore causes a permanent deformation in the shape of the earth, the equatorial diameter being slightly increased and the polar diameter slightly decreased.

Mean sea level also varies as $(3 \sin^2 f - 1)$ and is thus high when declination is great and low when declination is small.

§ 16. From expressions (18) and (19)

Mean half range of tide, or mean movement above or below mean sea level,

$$= \tfrac{3}{2}t \cos^2 l \cos^2 f, \tag{23}$$

and, from expression (5), the height varies as $\cos 2h$; the shape of the tide wave due to either the sun or moon is therefore a cosine, or harmonic, curve.

§ 17. The lunar and solar tide waves cannot exist independently; when two waves are superimposed the resultant wave is the sum of its two components, the height of the water at any moment being equal to the sum of the heights of the components at that moment.

If $\qquad \tfrac{3}{2}t \cos^2 l \cos^2 f$ of the moon $= m$

and $\qquad\qquad\qquad$ of the sun $\quad = s,$

and if $h =$ the moon's hour angle, and $H =$ the sun's hour angle, then at any moment the average height of the lunar tide $= m \cos 2h$, the average height of the solar tide $= s \cos 2H$, and

$$\text{Height of combined wave} = m \cos 2h + s \cos 2H. \tag{24}$$

§ 18. The length of a sidereal day, or period of the earth's rotation on its axis with reference to a fixed star, is about 23·9 hours.

Owing to the earth's motion round the sun, the length of a mean solar day is increased to 24 hours and the apparent angular velocity of the sun is $360°/24 = 15°$ per hour.

Owing to the moon's motion round the earth, the length of a mean lunar day is increased to about 24·8 hours and the apparent angular velocity of the moon is $360°/24·8 = 14·5°$, nearly, per hour.

Were the angular velocities of the sun and moon equal the angular distance of the moon from the sun would be constant; the difference in the velocities is small and the angular distance may therefore be considered constant for a short period of time, and

Angular distance of moon from sun = Time of moon's transit = $H \pm h$. (25)

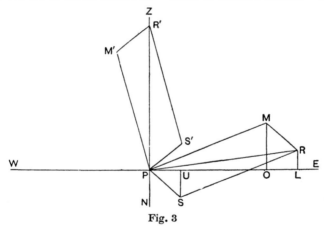

Fig. 3

§ 19. In Figure 3, $PM = PM' =$ mean half range of lunar tide $= m$, $PS = PS' =$ mean half range of solar tide $= s$; WE is mean sea level at P, ZN is perpendicular to WE; ZPM and $ZPM' =$ different values of $2h$, ZPS and $ZPS' =$ corresponding values of $2H$; $MPS = M'PS' = 2H \pm 2h$. MR is parallel to PS, SR parallel to PM, $M'R'$ parallel to PS' and $R'S'$ parallel to $M'P$; RL, MO and SU are perpendicular to WE.

Let $H \pm h = \theta$, then MPS and $M'PS' = 2\theta$.

$$MO = m \cos PMO = m \cos 2h. \quad (26)$$

$$SU = s \cos PSU = -s \cos 2H. \quad (27)$$

$$RL = MO - SU = m \cos 2h + s \cos 2H = \text{Height of tide}. \quad (28)$$

High water will occur when RL reaches its maximum value above WE.

$$RL = PR \cos PRL = \sqrt{m^2 + s^2 + 2ms \cos 2\theta} \cos ZPR. \quad (29)$$

RL therefore reaches its maximum value when $ZPR = 0$ or when RL coincides with RP and lies on ZN at $R'P$.

Height of high water above mean sea level $= R'P$

$$= \sqrt{m^2 + s^2 + 2ms \cos 2\theta}, \tag{30}$$

$M'PR' = 2h$ at time of high water

$\quad\quad = 2$ Interval between moon's transit and high water

$$= \tan^{-1} \frac{s \sin 2\theta}{m + s \cos 2\theta}.$$

Therefore:

Interval between moon's transit and high water

$$= \tfrac{1}{2} \tan^{-1} \frac{s \sin 2\theta}{m + s \cos 2\theta}, \tag{31}$$

$R'PR = 2$ Interval from high water when tide is at height RL.

Therefore, from expression (29)

Height of tide at any time

$$= \cos 2 \text{ Interval from high water } \sqrt{m^2 + s^2 + 2ms \cos 2\theta}. \tag{32}$$

From expressions (30) to (32), which it must be remembered include only the semi-diurnal portion, see expression (5), of the lunar and solar tides:

Height of low water below mean sea level

$$= -\sqrt{m^2 + s^2 + 2ms \cos 2\theta}, \tag{33}$$

$$\text{Time of high water} = \theta - \tfrac{1}{2} \tan^{-1} \frac{s \sin 2\theta}{m + s \cos 2\theta}, \tag{34}$$

$$\text{Time of low water} = 90° + \theta - \tfrac{1}{2} \tan^{-1} \frac{s \sin 2\theta}{m + s \cos 2\theta}, \tag{35}$$

and   (a) At new and full moon, when $\theta = 0$ or 12 hours:

High water occurs at the time of the moon's transit and reaches its maximum height of $(m + s)$ above mean sea level and low water its maximum height of $(m + s)$ below mean sea level; these are called *Spring* tides.

(b) At quadrature, when $\theta = 6$ or 18 hours:

High water occurs at the time of the moon's transit and reaches its minimum height of $(m - s)$ above mean sea level and low water its minimum height of $(m - s)$ below mean sea level; these are called *Neap* tides.

(c) When θ is between 0 and 6 hours or between 12 and 18 hours:

High water occurs before the time of the moon's transit, and the tide is said to *prime*.

(d) When θ is between 6 and 12 hours or between 18 and 0 hours:

High water occurs after the time of the moon's transit, and the tide is said to *lag*.

Note.—It must be remembered that, in §16 to §19, only the semi-diurnal portion of the tide wave is considered, and, further, that the whole of Chapter I refers to the tides which would be found on the perfect tidal earth (see §3), not to the tides as they exist.

CHAPTER II

The Tides as they exist

§ 20. *A free wave* is one which, once produced, would, apart from friction between particles, continue for ever; a *forced wave* is one due to the continuous application of external force.

The speed of a free wave, which depends on its length and the depth of water, may be calculated by complicated mathematical formulae, but in deep water, that is to say, where the depth exceeds the length of the wave, the speed depends on length only, and in shallow water, or where the depth is only a small fraction of the length, speed depends on depth only.

In shallow water the speed of a free wave varies as the square root of the depth, and in a depth of 11 yards speed is 20 miles per hour.

The speed of a forced wave depends only on the force.

§ 21. In Chapter I the tide wave was shown to be formed by the combination of two waves, due to the moon and sun respectively, and the lunar and solar waves were shown to contain semi-diurnal and diurnal portions; the combined wave at any place may therefore be considered, apart from changes due to changes in declination and parallax, as compounded of four unvarying waves, called *partial waves*, the height of the compound wave at any place and at any moment being the sum of the heights of the partial waves at that place and moment, and high water occurring when this sum reaches its maximum value above mean sea level; but as the relative positions of the moon and sun vary from day to day, so also do the relative positions, but not the magnitudes, of the partial waves vary, and consequently the height of the tide at a certain time, and the time and height of high water, vary from day to day.

The length of the semi-diurnal lunar and solar partial waves is half the circumference of the earth, or about 12,000 miles on the equator and 6000 miles in latitude 60°; the length of the diurnal lunar and solar partial waves is the circumference of the earth, or about 24,000 miles on the equator and 12,000 miles in latitude 60°. The interval between high waters of the solar semi-diurnal partial wave is 12 hours and between high waters of the solar diurnal partial wave 24 hours, the speed of both solar waves is therefore about 1000 miles per hour on the equator and 500 miles per hour in latitude 60°; the interval between high waters of the lunar semi-diurnal partial wave is 12·4 hours and between high waters of the lunar diurnal partial wave 24·8 hours, the corresponding speed of both lunar partial waves is therefore about 900 miles per hour on the equator and 450 miles per hour in latitude 60°.

§ 22. On the earth's surface the water area consists of the Southern ocean, completely surrounding the earth between the Antarctic continent, Australia and the southern extremities of Africa and South America, the Atlantic and Pacific oceans, forming gulfs extending to northward and joined at their northern extremes by the Arctic ocean, and the Indian ocean, forming a gulf extending to northward to a little past the equator.

The compound tide wave, as described in Chapter I, is formed by a change in shape of the fluid exterior of the earth; it is evident that no change of this nature can take place except where the earth is entirely surrounded by water, and the complete change therefore occurs in the Southern ocean only, and there only are the four partial waves formed; these combine to form the Southern ocean tide wave.

§ 23. As the depth of water is nowhere more than a very small fraction of the length of any partial wave, all are shallow water waves; in latitude 60° S., which may be considered to be the middle latitude of the Southern ocean, speeds are 500 miles per hour for the solar partial waves and 450 miles per hour for the lunar partial waves; the speed of a free shallow water wave is, as stated at § 20 above, 20 miles per hour in a depth of 11 yards and speed varies as the square root of the depth, therefore:

The speed of a free wave would be 500 miles per hour in a depth of

$$\frac{500^2 \times \sqrt{11}}{20^2} \text{ yards} = 3\cdot9 \text{ miles}, \tag{36}$$

and

The speed of a free wave would be 450 miles per hour in a depth of

$$\frac{450^2 \times \sqrt{11}}{20^2} \text{ yards} = 3\cdot3 \text{ miles}. \tag{37}$$

The solar partial waves can therefore not be free waves in latitude 60° S. except in depths of 3·9 miles and the lunar partial waves cannot be free waves except in depths of 3·3 miles.

As these critical depths exist, if at all, only in isolated spots, both lunar and solar partial waves in the Southern ocean are forced waves.

In water of less than the critical depth the speed of a free wave is less than the speed of the forced waves, in water of more than the critical depth the speed of a free wave is greater than that of the forced waves, therefore as the depth decreases from the critical depths the forced partial waves lag behind the sun and moon, as the depth increases from the critical depths the forced partial waves get ahead of the sun and moon. The tide raising forces of a body, either the sun or moon, are at their greatest at the transit, either upper or lower, of the body, which therefore tends to cause high water at places where it is on the meridian; the depth of water tends either to decrease or to increase the speed of the partial waves, and therefore to retard or advance the time of high water, and the position of the crest of each partial wave varies according to the resultant of these tendencies; as however, at any place, the depth of water is fixed and unalterable, so also at that place, apart from changes due to changes in declination and parallax, is the position of the crest of each partial wave fixed with reference to the body. As the depth of water is, almost without exception, less than the critical depth, the crest is always behind the body, and high water of each partial wave in the Southern ocean therefore always occurs after the transit of the body.

§ 24. Change of surface level in the Southern ocean must, of necessity, cause changes in the gulfs formed by the Atlantic, Pacific and Indian oceans, and the Southern ocean forced partial waves accordingly generate secondary waves which penetrate the gulfs in a northerly direction and cause tides in those waters.

The secondary waves are free waves, for, after parting from the Southern ocean forced waves, they are under no constraint and progress according to the law of shallow water free waves, their speed increasing or decreasing as the depth increases or decreases. Each wave always occupies the same time in traversing the same space, the law as to time will therefore be the same as for the Southern ocean forced waves, and, apart from changes due to changes in declination and parallax, the crest of each solar partial wave will reach a certain place at a definite and fixed interval after the transit of the sun, the crest of each lunar partial wave at a definite and fixed interval after the transit of the moon, but, as the waves progress to northward, the intervals will increase to northward.

§ 25. The secondary waves cannot affect enclosed seas, such as the Mediterranean, and can only affect partially enclosed seas, such as the Gulf of Mexico, to a very small extent; the tides in such seas must therefore be locally developed.

The tide developed as in Chapter I is, at any moment, either above mean sea level or below mean sea level over a very large area; the quantity of water in an enclosed sea on any day is invariable and tides of this nature are therefore impossible.

The vertical component of the moon's tide-raising force can have no tidal effect on enclosed seas, for, from expression (4),

$$\text{Force of vertical component} = \frac{M}{E} \left(\frac{r}{d}\right)^3 g \, (3 \cos^2 \alpha - 1)$$

$$= 2\frac{M}{E} \left(\frac{r}{d}\right)^3 g \text{ when } \alpha = 0°.$$

$$= -\frac{M}{E} \left(\frac{r}{d}\right)^3 g \text{ when } \alpha = 90°.$$

Therefore the greatest effect of the vertical component, which will occur during the change in α from $0°$ to $90°$, $= 3 \dfrac{M}{E} \left(\dfrac{r}{d}\right)^3 g$

$$= 0{\cdot}00000025g.$$

The greatest change in the level of the surface of an enclosed sea due to the vertical component of the moon's tide-raising force can therefore not exceed $0{\cdot}00000025$ of its depth, or about $0{\cdot}0013$ ft. for every mile of depth; the vertical component of the sun's tide-raising force may increase this to about $0{\cdot}002$ ft. per mile of depth, but this quantity would be quite inappreciable in any existing sea.

The horizontal component of the moon's tide-raising force has not been considered; in Figure 2 (p. 3) $JPK = (\alpha + \beta)$, by construction $\tan \beta = \frac{1}{2} \tan \alpha$, and PK, the moon's tide-raising force, represents a force of

$$\mu M \left(\frac{1}{D^2} - \frac{1}{d^2}\right) \frac{1}{\cos \beta}.$$

Therefore JK, the horizontal component of the moon's tide-raising force, represents a force of

$$\mu M \left(\frac{1}{D^2} - \frac{1}{d^2}\right) \frac{\cos (\alpha + \beta)}{\cos \beta}$$

$$= \frac{3}{2} \frac{M}{E} \left(\frac{r}{d}\right)^3 g \sin 2\alpha. \tag{38}$$

The value of α may be considered known, for

$$\cos \alpha = \cos l \cos f \cos h \pm \sin l \sin f$$

and the horizontal component of the moon's tide-raising force is therefore known.

This force, acting horizontally, causes a small displacement of the plumb line and therefore, as the surface of the sea remains perpendicular to the plumb line, of sea surface. The greatest displacement of the plumb line occurs when α changes from $45° +$ to $45° -$, or, on the earth's equator and when declination is $0°$, when the moon's hour angle changes from $45°$ east to $45°$ west.

In Figure 4, P is a place on the surface of a sea extending in an east and west direction, EW is the mean position of the surface and RP, the mean direction of the plumb line, is perpendicular to EW.

If RP represents, both in direction and magnitude, the undisturbed force, g, of gravity, and RH and RV represent disturbing forces of $\dfrac{3}{2} \dfrac{M}{E} \left(\dfrac{r}{d}\right)^3 g$ acting east and west respectively, then HP will represent the direction of the plumb line and OD the surface of the sea when the easterly force is acting, VP will represent the direction of the plumb line and NU the surface of the sea when the westerly force is acting.

UD therefore represents the surface displacement, or tide, at E due to the horizontal component of the moon's tide-raising force acting at P.

$$UD = 2UE = 2PE \tan UPE$$
$$= 2PE \tan RPH$$
$$= 3PE \, \frac{M}{E} \left(\frac{r}{d}\right)^3$$
$$= 0{\cdot}0000002PE$$
$$= 0{\cdot}5 \text{ ft. when } PE \text{ is 500 miles.} \tag{39}$$

As the sun's tide-raising force is about half that of the moon, the greatest tide, due to the horizontal component of the moon's and sun's tide-raising forces, in a distance of 500 miles is about 0·75 ft.

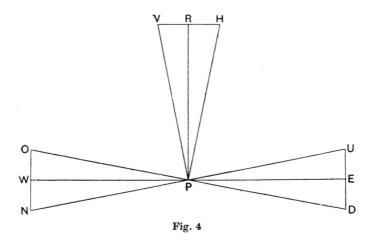

Fig. 4

The real effect of the force is not easily determined, for the moon and sun act on all points of the surface of the sea, not on one point only, but it is evident that an oscillatory tide will be set up, that this will take the form of forced waves travelling from east to west, and that, as the waves vary with sin 2α, they will have semi-diurnal and diurnal portions as in the Southern ocean forced waves and the Atlantic, Pacific and Indian ocean free waves. East to west forced waves of a similar nature must also exist in the oceanic gulfs wherever the width is sufficient to admit of their development, but both in enclosed seas and in the oceanic gulfs each wave so formed will, apart from changes due to changes in declination and parallax, reach any place at a definite and fixed interval after the transit of the sun or moon. It is also evident that the free waves in the oceanic gulfs will be complicated by the east to west forced waves.

§ 26. Neither the absolute nor relative magnitudes of the partial waves as they exist are in agreement with magnitudes derived from expressions (5) and (39), and consideration shows that these could not be maintained unless the earth were entirely covered by water of uniform depth, for:

(*a*) The energy contained in a free wave varies with its length, height, width and speed and, apart from loss by friction, the energy remains constant; if, therefore, the width is decreased by converging shore lines whilst the length and speed remain constant the height must be increased, and if the speed is decreased owing to shoaling water whilst the length and width remain constant the height must also be increased. Converging shore lines and decreasing depth thus increase height, diverging shore lines and increasing depth decrease height.

(*b*) Where the progress of the partial waves is obstructed by reefs and shoals, the long period diurnal partial waves have more time for overcoming the obstructions than the shorter period semi-diurnal partial waves and therefore reach the farther side in a more complete condition than the shorter period semi-diurnal partial waves.

(*c*) Where the direction of progress of the partial waves is not parallel to the shore line secondary waves are reflected from the shore; these affect the diurnal and semi-diurnal partial waves to a different extent.

(*d*) When an advancing wave is divided by land and the divisions reunite after passing the land, one division may experience delay, due to shallow water, not experienced by the other. Under these conditions, if the delay amounts to 6 hours high water and low water of the semi-diurnal partial waves will meet and there may be an area in which the semi-diurnal tide is non-existent, if the delay amounts to 12 hours high water and low water of the diurnal partial waves will meet and there may be an area in which the diurnal tide is non-existent, if the delay is between 0 hour and 6 hours or between 6 hours and 12 hours double high and low waters, or *double half day tides*, may occur.

§ 27. Though the speed of a shallow water free wave varies with the depth of water, this rule is only true when the progress of the wave is uninterrupted, and reflected waves and obstructions due to land and shoals affect the speed of the partial waves equally with their height, the short period waves being retarded to a greater extent than those of long period; the solar partial waves are therefore more retarded than the lunar partial waves; high water of the lunar diurnal partial wave does not necessarily occur at the same time as high water of the lunar semi-diurnal partial wave, and high water of the solar semi-diurnal partial wave does not necessarily occur at the same time as high water of the solar diurnal partial wave.

§ 28. The tide wave as it exists at all places is thus formed by the combination of a number of partial waves, some of which are forced waves directly due to the tide-raising forces of the moon and sun, others are free secondary waves derived from the forced waves and others again reflected waves, but both the free secondary waves and the reflected waves are indirectly due to the tide-raising forces of the moon and sun; some partial waves make high water twice a day, others once a day only. The lunar partial waves vary with changes in the moon's declination and parallax, the solar partial waves with changes in the sun's declination and parallax, but, apart from such changes, the magnitude of each partial wave at a place is constant and each partial wave reaches the place at a constant interval after the transit of the body to the tide-raising force of which it was originally due. The height of the compound tide wave at a place at any moment is the sum of the heights of all the partial waves at that place and moment; high water therefore occurs when this sum reaches its maximum height, low water when it reaches its minimum height. As the magnitudes of the partial waves vary with changes in the moon's and sun's declination and parallax, and as the interval between the arrival of the crests of lunar and solar partial waves varies with changes in the angular distance between the moon and sun, the times and heights of high and low water of the compound wave also vary with changes in the moon's and sun's declination and parallax and with their angular distance, which may be measured by the time of the moon's transit.

On the ideal tidal earth semi-diurnal tides would occur at the equator, diurnal (single day) tides at a distance from the equator depending on the declinations of the moon and sun (see expression (5) and §14, *Note*); high water of each semi-diurnal partial wave would occur at the transit, upper or lower, of the tide-raising body, high water of each diurnal partial wave at its upper transit, low water of each semi-diurnal partial wave when the body was on the horizon and low water of each diurnal partial wave at its lower transit. On the earth as it exists the tidal changes occur which would occur under ideal conditions, but the arrival of each partial wave is delayed, changes are delayed and the absolute and relative magnitudes of the partial waves and changes are profoundly modified by terrestrial conditions.

§ 29. The tide at any place on the earth's surface may consequently be:

(*a*) Almost entirely semi-diurnal, that is to say, owing to increase in the magnitude of the semi-diurnal, and decrease in the magnitude of the diurnal partial waves, there may be two nearly equal tides in each lunar day. Where semi-diurnal tides occur the most noticeable variation is from springs to neaps and *vice versa*, high water and low water, springs and neaps, priming and lagging occur in, or nearly in, accordance with expressions (30) to (35) and § 19 (*a*) to (*d*), changes due to changes in declination and parallax occur in, or nearly in, accordance with expression (5), but all phenomena and changes are delayed and occur later than they would occur under ideal conditions.

(*b*) Almost entirely diurnal, that is to say, owing to increase in the magnitude of the diurnal, and decrease in the magnitude of the semi-diurnal partial waves, there may be only one tide in each lunar day. Where diurnal tides occur the most noticeable variation follows changes in the moon's declination and, though all other changes and phenomena occur as at (a) above, these may be so small as to be invisible on superficial examination.

(c) Of any type between the extremes described at (a) and (b); it may even happen that the most noticeable variation follows changes in the moon's parallax instead of in the angular distance between the moon and sun or in the moon's declination. Double high and low waters may occur with tides of any type.

Thus, though the tide may appear to be of entirely different type at different places, all types are essentially the same, differences being entirely due to changes in the absolute and relative magnitudes of the partial waves, to changes in the absolute and relative magnitudes of the variations due to changes in declination and parallax, and to delays due to terrestrial conditions.

§ 30. Though in Europe the tide is usually of semi-diurnal type, as described at § 29 (a), single day tides, § 29 (b), occur in the Baltic and double half day tides in parts of the English Channel and on the Netherlands coast, but in other parts of the world single day tides are common and, as the transition is gradual, all types, from the semi-diurnal with hardly a trace of diurnal, as for instance at London, to the diurnal with hardly a trace of semi-diurnal, as for instance at Do Son, French Indo-China, are found.

In Figures* 5 and 6 tides which approach the extremes, § 29 (a) and (b), are shown. Figure 5 gives a typical 15 day curve for Dover, where the tide is almost entirely semi-diurnal, § 29 (a); in this figure the maximum range, springs, occurs on about the 2nd and 15th days, though the range on the 15th day, when the tide falls below chart datum (see § 34), is considerably greater than that on the 2nd day; the minimum range, neaps, occurs on about the 9th day. At the commencement and end of the figure there is no appreciable diurnal tide; in the middle of the figure the diurnal tide, though small, is at its maximum. It may thus be assumed from the figure that new or full moon occurred near the 2nd and 15th days and quadrature near the 9th day; that the moon was in apogee near the 2nd day and in perigee near the 15th day, and that the moon's declination was zero near the 1st and 15th days and maximum near the 9th day.

Figure 6 gives a typical 15 day curve for Victoria, B.C., where the tide is almost entirely diurnal, § 29 (b); in this figure, on the 1st day the range of the tide is comparatively small and there is only one true high and one true low water, though the rise to high water is interrupted by a slight fall which indicates the second high and low water. From the 1st day to about the 8th day the range of the diurnal tide increases, then decreases to the 15th day; the semi-diurnal tide vanishes on the 2nd and 3rd days, then increases to a maximum, which is still very small, on about the 9th day, then again decreases and is evidently about to vanish on the 15th day. It may thus be

* See folding plate at end of book.

assumed from the figure that the moon's declination was zero near the 1st and 15th days and maximum near the 8th day, also that new or full moon occurred near the 9th day and quadrature near the 2nd and 15th days.

The conclusions reached from the figures are not final, for the sun affects the tides and the age (see § 38) must also be considered, but the figures show that, whether the tide is semi-diurnal or diurnal, it is compounded of the same partial waves and that the same astronomical conditions have the same effects, though both partial waves and astronomical effects differ in absolute and relative magnitude at different places.

The curves in the figures were not obtained from observed tides, but were predicted from the partial waves by means of the tide predicting machine (Chapters VI and VII), and the fact that such essentially different tides can be predicted from partial waves which, though differing in absolute and relative magnitude, are the same in both cases, shows conclusively that the tide wave as it exists is a compound, not a simple, wave.

CHAPTER III

Definitions. Approximate Tidal Prediction by means of Non-harmonic Tidal Constants

§ 31. Certain definitions, required for explaining the cause of tides, were given in Chapters I and II; these are not all suited to the tides as they exist and the phenomena defined are therefore more closely described below.

Tidal problems may, broadly speaking, be dealt with by two methods—harmonic and non-harmonic. A harmonic wave is one represented by the formula

$$y = \tfrac{1}{2}r \cos h, \tag{40}$$

where y is the difference between sea level and mean sea level, r the range of the wave, and h the interval, or proportion of the half-period of the wave expressed in degrees, between the time of high water and the time at which the height is measured; in the harmonic method, the compound tide wave is considered to be formed by the super-position of a number of partial waves, each of harmonic form, arranged in such a manner as to represent not only the semi-diurnal and diurnal partial waves described in Chapter II, but also the changes in the tide due to changes in moon's and sun's declination and parallax. In the non-harmonic method the compound wave is considered to be formed by the solar and lunar semi-diurnal partial waves only, the diurnal partial waves and changes due to changes in lunar and solar declination and parallax being treated as corrections to the compound wave in exact prediction and being neglected in approximate prediction.

The harmonic method must be used for predicting the tides in regions where the diurnal partial waves are large in proportion to the semi-diurnal; in these regions, if the sum of the ranges of the lunar and solar diurnal partial waves exceeds the difference between the ranges of the lunar and solar semi-diurnal partial waves, one high water of the semi-diurnal compound wave will, at times, be entirely masked by low water of the diurnal compound wave, and single day tides will occur, with only one high water and one low water in a lunar day; it is evident that, in such regions, the compound wave cannot be even approximately represented by means of the semi-diurnal partial waves only.

The non-harmonic method may be used in regions where the diurnal partial waves are small in proportion to the semi-diurnal; as this is generally the case in European waters, and as it is the method in which the information given in the Admiralty Tide Tables, Part II, must be used, it will be first described.

Though the method is called non-harmonic, the semi-diurnal partial waves and the compound wave are assumed to be of harmonic form; both methods are therefore in reality harmonic, the difference being that in the so-called harmonic method the partial waves themselves are dealt with whereas in the so-called non-harmonic method the compound wave is considered as a whole; approximate predictions by this method should only be used when information for calculating more exact predictions is not available (see Chapters **IV** and **VII**).

§ **32.** *Tidal constants* are the intervals and heights used for describing the tide; though they are unvarying at any place they do not always appear so on superficial examination. Constants may be either harmonic or non-harmonic, harmonic constants referring to the partial waves, non-harmonic constants to the compound wave.

A Standard port is a port for which the times and heights of high and low water are completely and accurately calculated and published in the Admiralty Tide Tables, Part I.

A Secondary port is a port for which tidal differences on a Standard port are published in the Admiralty Tide Tables, Part II.

Tidal differences are quantities, either of time or of height, which, when applied to the accurately predicted times or heights of the tide at a Standard port, give approximately the corresponding times or heights at a Secondary port.

§ **33.** *Mean sea level* is the average level of the sea as calculated from a long series of observations obtained at equal intervals of time.

Mean tide level is the average level of the sea as calculated from a long series of observations of the heights of high and low water.

There is, as a rule, a small difference between mean sea level and mean tide level; were the compound tide wave of exact harmonic form this difference would not exist, but as the wave is seldom, or never, exactly harmonic, the average height (mean sea level) does not agree exactly with the mean of the extremes (mean tide level).

Neither mean sea level nor mean tide level is constant, for both are affected by astronomical and meteorological conditions and also by changes of long period; such effects must be considered in accurate tidal predictions, but may be ignored in approximate predictions, and for this purpose *mean level* is assumed to be a fixed point about which the tide oscillates, and mean level is thus a tidal constant (this assumed fixed point is called mean tide level in the Admiralty Tide Tables, Part II).

§ **34.** Mean level is not a suitable reference level for soundings on charts, for the sea surface is below it as frequently as above it, and an arbitrary plane of reference, called *chart datum*, is therefore adopted.

Chart datum should be a plane so low that the tide will not often fall below it; it is sometimes said to be the level of the lowest possible low water, of mean low water springs, of mean low water, etc., but is in the large majority of cases an arbitrary low water level and is best so considered in every case.

In most civilized countries levels are referred to a fixed land survey datum which has, by a system of levelling, been carried to all parts of the country. The land survey datum is also sometimes said to be, or to be referred to, mean sea level, mean low water spring level, etc., but as these levels had seldom been accurately calculated when the land survey datum was established, and as they are, in any case, not constant, the land survey datum should also always be considered as an arbitrary level.

The land survey datum in the United Kingdom is Ordnance datum; in Great Britain Ordnance datum is said to be mean sea level at Liverpool, but is, in reality, 4·67 ft. above the level of the sill of the Old dock at that port; this dock is no longer in existence but the level of the sill is preserved on the river face of the centre pier of Canning half-tide dock; average mean sea level at Liverpool, calculated from 8 years' tidal observations, is 0·34 ft. above Ordnance datum. Ordnance datum in Ireland is said to be the level of mean low water springs in Dublin Bay, but is, in reality, 20·90 ft. below a bench mark cut in the base course, under the south window, of Poolbeg lighthouse; the true average level of mean low water springs in Dublin Bay, as calculated from 2 years' observations, is 2·59 ft. above Ordnance datum.

Chart datum should be connected with the land survey datum and with the best obtainable value of mean level; in countries where no land survey datum exists chart datum should be connected with some permanent prominent fixed mark. As the range of the tide differs at all places, and as chart datum at any place depends, to some extent, on the range, chart datum also differs at all places, but the relative level at different places may be ascertained by means of the connection with land survey datum and mean level.

§ 35. The tide rises to a level called *high water* and falls to a level called *low water*, the difference between these levels is the *range* of the tide; the difference between the levels of high water and of chart datum is the *rise* of the tide; the difference at any moment between sea level and chart datum is the *height* of the tide.

As the tide oscillates about mean level, the difference between high water level and mean level is equal to the difference between mean level and low water level, and is the *half*, or *semi*, *range* of the tide.

The tide does not commence to fall immediately it ceases to rise, neither does it commence to rise immediately it ceases to fall; the interval between the cessation of rising and the commencement of falling is the *high water stand* of the tide, and the middle moment of the high water stand is the *time of high water*; similarly the interval between the cessation of falling and the commencement of rising is the *low water stand* of the tide, and the middle moment of the low water stand is the *time of low water*.

§ 36. An *Inequality* in the tide is any deviation, either in time or in height, from the average; where the semi-diurnal partial waves predominate the average tide is the lunar semi-diurnal partial wave.

Inequalities are distinguished either by their causes or their effects, and are:

(*a*) The *Phase Inequality* giving the effect on the lunar semi-diurnal partial wave of the solar semi-diurnal partial wave; this inequality varies with the angular distance between the sun and moon, and therefore with the time of the moon's transit, and is the only inequality considered in approximate prediction by non-harmonic methods.

(*b*) The lunar and solar *anomalistic inequalities*, due to variations in the moon's and sun's horizontal parallax, or distance from the earth.

(*c*) The lunar and solar *declinational inequalities*, due to variations in the moon's and sun's declination.

(*d*) The lunar and solar *diurnal inequalities*, due to the lunar and solar diurnal partial waves, and varying with the moon's and sun's declination.

All inequalities are considered in exact predictions by non-harmonic methods.

§ 37. The following abbreviations are useful:

H.W = high water	L.W. = low water
S. = spring tides	N. = neap tides
T. = tide	M. = mean
L. = level	I. = interval
F. & C. = full and change (of moon).	

Two or more abbreviations may be used together, thus:

M.H.W.S. = mean high water springs
M.L. = mean level
M.L.W.N. = mean low water neaps
etc., etc.

§ 38. In the approximate predictions now to be explained the lunar and solar semi-diurnal partial waves only are dealt with, changes due to changes in declination and parallax are not considered, and each partial wave, at any place, is therefore of constant magnitude and reaches the place at a constant interval after the transit of the tide-raising body. As the sun crosses the meridian of a place at intervals of 12 hours, the interval between the sun's transit and high water of the solar semi-diurnal partial wave cannot exceed 12 hours; similarly the interval between the moon's transit and high water of the lunar semi-diurnal partial wave cannot exceed 12 hours of lunar time or 12·4 hours of solar time; as the sun is on the meridian at noon or midnight, if S be the interval between the sun's transit and high water of the solar semi-diurnal partial wave, S will also be the time of high water of the solar semi-diurnal partial wave, but if M be the interval between the moon's transit and high water of the lunar semi-diurnal partial wave and θ be the time of moon's transit, $\theta + M$ will be the time of high water of the lunar semi-diurnal partial wave.

When the moon is on the meridian of a place the angular distance between the sun and moon is θ but the angular distance between high waters of the solar and lunar semi-diurnal partial waves is $(\theta + M) - S$; M and S are constant, θ changes 0·8 hour in each lunar day, therefore $\dfrac{M - S}{0\cdot8}$ days from the time of high water $(\theta + M)$ the time of the moon's transit was $(\theta + M) - S$.

It was stated in § 27 that, though all partial waves are retarded, the retard of a short period wave is greater than that of one of longer period, the solar partial wave is therefore retarded to a greater extent than the lunar, $\dfrac{M - S}{0\cdot8}$ is a minus quantity and the phase inequality does not depend on the time of the moon's transit θ on the day, as in expression (34), but on a time of transit $(\theta + M) - S = \theta'$ which occurred $\dfrac{M - S}{0\cdot8}$ days earlier; this interval is called the *Age** of the tide, and

$$\text{Time of high water} = \theta + M - \tfrac{1}{2}\tan^{-1}\frac{s\sin 2\theta'}{m + s\cos 2\theta'}. \tag{41}$$

The age of the tide is sometimes, though incorrectly, explained as being the time occupied by the compound wave in travelling up the oceanic gulfs; this is not a true explanation, for the ratio between the solar and the lunar semi-diurnal partial waves, though constant at a particular place, differs at different places on the same ocean, and the partial waves are not equally delayed; it is evident that if the ratio between the component parts of the compound wave changes, and if the component parts travel at different speeds, the compound wave cannot be considered to travel as a single wave.

M is the interval between the moon's transit and high water of the lunar semi-diurnal partial wave, but as, for all values of θ, the phase inequality disappears, M is also the average interval between the moon's transit and high water of the compound wave, and is therefore called the *mean high water lunitidal interval* (M.H.W.I.) or *mean establishment*.

On the ideal tidal earth considered in Chapter I the time occupied by the tide either in rising or in falling is, apart from the phase inequality, always one quarter lunar day exactly; on the earth as it exists the period occupied in rising frequently differs considerably from that occupied in falling, and an interval M', the *mean low water lunitidal interval* (M.L.W.I.), is therefore required for finding the time of low water; this interval is, as a matter of convenience,

---

* The age of the tide is, strictly speaking, the interval from the time of moon's transit θ' to the time of high water $\theta + M$, but, as a matter of convenience, the interval M is subtracted from this quantity and the age considered to be the number of half lunar days between transits θ' and θ.

always taken to be that of the low water following the high water given by the interval M, and

$$\text{Time of low water} = \theta + M' - \tfrac{1}{2}\tan^{-1}\frac{s\sin 2\theta'}{m + s\cos 2\theta'}. \tag{42}$$

Similarly, if A be the height of mean level above chart datum, the heights of high and low water above chart datum will not be given by expressions (30) and (33), but

$$\text{Height of high water} = A + \sqrt{m^2 + s^2 + 2ms\cos 2\theta'}, \tag{43}$$

$$\text{Height of low water} = A - \sqrt{m^2 + s^2 + 2ms\cos 2\theta'}. \tag{44}$$

As a consequence of expressions (41) to (44), § 19 (a) to (d) requires amendment as follows:

(a) Spring tides occur at an interval equal to the age of the tide after new and full moon.

(b) Neap tides occur at an interval equal to the age of the tide after the moon's quadrature.

(c) The tides prime from springs to neaps.

(d) The tides lag from neaps to springs.

And further:

(e) The H.W.I. and L.W.I. on the days of spring and neap tides are the M.H.W.I. and M.L.W.I., i.e. on those days there is no phase inequality.

Though from expressions (43) and (44) it would appear that the range of the tide is always greatest at springs and least at neaps, this is not the case in reality, for these expressions include only the average semi-diurnal partial waves, the diurnal, anomalistic and declinational inequalities being omitted; it is evident that where the sum of these inequalities exceeds the phase inequality the range at neaps may exceed the range at springs, and that even where the neglected inequalities are small the range a day or two before or after springs may exceed the spring range and the range a day or two before or after neaps may be less than the neap range.

§ 39. Expressions have now been found by means of which the times and heights of high and low water may be approximately calculated provided that the neglected inequalities are small and that:

The time of moon's transit,

The age of the tide,

The M.H.W.I. and M.L.W.I.,

The semi-ranges m and s, and A, the height of mean level above chart datum, are known.

The times and heights of high and low water are, however, not usually calculated by means of these expressions, though there is no reason why they should not be, and it is in fact desirable that they should be, for, as the expressions are harmonic, the difference between non-harmonic and harmonic methods of calculating approximate predictions (see Chapter VII) would then disappear, but the necessary constants are not given in the Admiralty Tide Tables.

§ 40. The H.W.I. at a certain place on any day is the M.H.W.I. plus the phase inequality, the L.W.I. is the M.L.W.I. plus the phase inequality. The phase inequality on any day depends on the time of the moon's transit on a day the age of the tide previously, and the ratio $m : s$; the time of moon's transit on a day the age of the tide previously depends on the time of transit on the day; the ratio $m : s$ does not change, and therefore, at any place, the phase inequality varies only with the time of the moon's transit and is constant for any particular time of transit; as the M.H.W.I. is also constant, the interval between the moon's transit and high water is constant for any particular time of transit.

On the days of new and full moon, when the moon is on the meridian at noon or midnight, the high water lunitidal interval is the time of high water, and this interval, called the *high water full and change* (H.W.F. & C.) or *vulgar establishment*, is the tidal constant given in the Admiralty Tide Tables, Part II, for finding the time of high water. The low water lunitidal interval on new and full moon days is similarly the time of low water on those days, and this interval, called the *low water full and change* (L.W.F. & C.), is the tidal constant given in the Admiralty Tide Tables, Part II, for finding the time of low water.

As the H.W.F. & C. and L.W.F. & C. include phase inequality, the inequality to be applied to these intervals when finding the times of high and low water is not the same as that to be applied to the M.H.W.I. and M.L.W.I.; the amount of phase inequality included in the H.W.F. & C. and L.W.F. & C. varies with the ratio $m : s$ and the age of the tide. The age of the tide is approximately known for all parts of the world and is given in the Admiralty Tide Tables, Part II, Table I; if therefore a universal ratio $m : s$ be assumed, both the amount included and the amount to be applied can be calculated beforehand and tabulated according to the time of moon's transit and age of the tide.

It has been found that a universal assumed ratio of $s : m = 1 : 2 \cdot 73$* gives an average result of sufficient accuracy for approximately predicting the times of high and low water at ports where all inequalities except the phase inequality are small; this ratio has accordingly been adopted and Table II (*a*) in the Admiralty Tide Tables, Part II, calculated as follows:

In Figure 7 the times of moon's transit are shown on the central horizontal line, the phase inequality calculated from expression (31) with the assumed ratio $s : m = 1 : 2 \cdot 73$ is shown by the curve, and is read off on the left vertical line.

It is evident that when the age of the tide is 0 day, M.H.W.I. = H.W.F. & C., and the phase inequality to be applied to either interval when age is 0 day, according to time of moon's transit, may therefore be obtained directly from the curve.

* The origin of this assumed ratio cannot be traced; a more accurate assumption would probably be $s : m = 1 : 2 \cdot 18$ (see § 12).

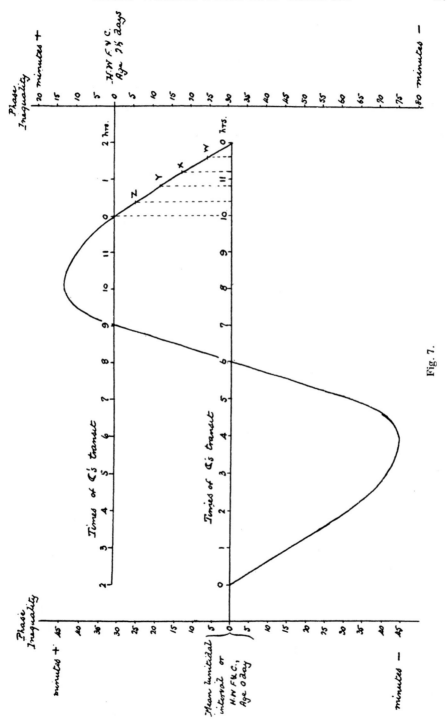

Fig. 7.

If the age of the tide is $2\frac{1}{2}$ days, then, when time of moon's transit is 0 hour, the phase inequality to be applied to the M.H.W.I. will be the inequality for moons' transit $0 - (0{\cdot}8 \times 2\frac{1}{2}) = 10$ hours, that is to say

$$\text{H.W.F. \& C.} = \text{M.H.W.I.} + \text{Phase inequality for transit 10 hours,} \qquad (45)$$

and the phase inequality to be applied to the H.W.F. & C., according to time of moon's transit when the age of the tide is $2\frac{1}{2}$ days, may therefore be found by graduating a horizontal line drawn through the point where the curve cuts a vertical line from time of moon's transit 10 hours, as shown in the figure, where also the amount of the phase inequality is shown on the right vertical line.

Were the age of the tide $\frac{1}{2}$, 1, $1\frac{1}{2}$ or 2 days, the phase inequality to be applied to the M.H.W.I. when time of moon's transit is 0 hour would be the inequality for moon's transit $11{\cdot}6$ hours, $11{\cdot}2$ hours, $10{\cdot}8$ hours, or $10{\cdot}4$ hours and the phase inequality to be applied to the H.W.F. & C. according to time of moon's transit would then be found by graduating horizontal lines through the points W, X, Y or Z on the curve.

The phase inequality to be applied to the L.W.F. & C. is assumed to be the same as that to be applied to the H.W.F. & C.

§ 41. The semi-ranges s and m of the solar and lunar semi-diurnal partial waves are constant at any place; A, the height of mean level at the place above chart datum, is also constant, therefore

$$\text{Mean spring rise} = A + (m + s), \qquad (46)$$

$$\text{Mean neap rise} \ \ = A + (m - s), \qquad (47)$$

are also constant, and the tidal constants for finding the heights of high and low water given in the Admiralty Tide Tables, Part II, are the heights of mean high water springs (M.H.W.S.), mean high water neaps (M.H.W.N.) and mean level (M.L.) above chart datum.

From expressions (46) and (47)

$$\text{Height of mean high water} = A + m, \qquad (48)$$

therefore, from expression (43),

$$\text{Phase inequality of height} = m - \sqrt{m^2 + s^2 + 2ms \cos 2\theta'}. \qquad (49)$$

Where both the spring rise and the neap rise are known a universal ratio $s : m$ must not be assumed; without this assumption the phase inequality given by expression (49) is not easily tabulated, but

$$\sqrt{m^2 + s^2 + 2ms \cos 2\theta'} = m + s \cos 2\theta' \text{ nearly,} \qquad (50)$$

therefore, from expressions (43) and (44),

$$\text{Height of high water} = A + (m + s \cos 2\theta'), \qquad (51)$$

$$\text{Height of low water} \ \ = A - (m + s \cos 2\theta'), \qquad (52)$$

m and s may be found from the constants by means of expressions (46) and (47); the time of moon's transit θ' varies with the time of transit θ and the age of the tide, and therefore with the interval between the day on which the height is required and spring tide, and the value of $\cos 2\theta'$ is given in Table II (b) in the Admiralty Tide Tables, Part II, according to the number of days from spring tide, which should be found as follows:

Days from spring tide

= Days from preceding new or full moon − age of tide,

or   = Days from following new or full moon + age of tide.   (53)

§ 42. Rules for finding approximately the times and heights of high and low water from the tidal constants given in the Admiralty Tide Tables, Part II.

(a) To find the time of high water:

To the time of the moon's transit add the H.W.F. & C. corrected for phase inequality.

(b) To find the time of low water:

To the time of the moon's transit add the L.W.F. & C. corrected for phase inequality.

The time of the moon's transit must be corrected for longitude if necessary, and the resulting times of high and low water will be in local mean time.

(c) To find the height of high water:

To half the sum of the heights of M.H.W.S. and M.H.W.N. apply, according to sign, half the difference between the heights of M.H.W.S. and M.H.W.N. multiplied by the phase inequality factor.

(d) To find the height of low water:

From twice the height of mean level subtract half the sum of the heights of M.H.W.S. and M.H.W.N. and apply to the result half the difference between the heights of M.H.W.S. and M.H.W.N. multiplied by the phase inequality factor with the sign changed.

As the low water found from the L.W.F. & C. follows the high water found from the H.W.F. & C., when the time and height of high water have been found, the time and height of the following low water may be found as follows:

(e) To find the time of low water:

To the time of high water add the difference between the H.W.F. & C. and the L.W.F. & C.

(f) To find the height of low water:

From twice the height of mean level subtract the height of high water.

In some cases, as tidal knowledge is still incomplete, certain constants are missing; where this occurs the following assumptions may be made:

L.W.F. & C.   = H.W.F. & C. + 6 h. 12 m.   (54)

Height of M.H.W.N. = $0 \cdot 75 \times$ Height of M.H.W.S.   (55)

Height of M.L.   = $0 \cdot 50 \times$ Height of M.H.W.S.   (56)

The assumed constants are not given in the Admiralty Tide Tables for the rule is to give no information which is not founded on observation and known to be at least approximately correct.

Information, etc. required for rules (*a*) to (*f*) will be found as follows:

Tidal constants for the port—Admiralty Tide Tables, Part II.

Time of moon's transit—Admiralty Tide Tables, Part I, Table IV (*a*), Nautical Almanac, or latest Supplement to Admiralty Tide Tables, Part II.

Days of new and full moon—Admiralty Tide Tables, Part I, Table IV (*b*), Nautical Almanac, or latest Supplement to Admiralty Tide Tables, Part II.

Age of tide—Admiralty Tide Tables, Part II, Table I.

Phase inequality of time—Admiralty Tide Tables, Part II, Table II (*a*).

Factor for phase inequality of height—Admiralty Tide Tables, Part II, Table II (*b*).

When the height of mean level above chart datum is given, the heights found by means of rules (*c*), (*d*) and (*f*) will be above chart datum; when the height of mean level above chart datum is not given and a height is assumed in accordance with expression (56), the heights found by means of the rules will be above an approximate mean low water spring level.

§ 43. *Example I.* What L.W.F. & C. and M.L. should be assumed at Whitby?

Admiralty Tide Tables, Part II, No. 218:

Whitby, H.W.F. & C. 3 h. 45 m., M.H.W.S. 15·0 ft.

Expression (54), L.W.F. & C. = 3 h. 45 m. + 6 h. 12 m. = 9 h. 57 m.

Expression (56), M.L. = 0·5 × 15·0 ft. = 7·5 ft.

Example II. Find the times and heights of the afternoon high and low waters at Grimsby on 3rd June, 1921.

Admiralty Tide Tables, Part I:

Table IV (*a*), 3rd June, 1921, moon's transit 9 h. 25 m.

Table IV (*b*), new moon, 6th June, 1921.

Admiralty Tide Tables, Part II:

Table I, age of tide, 2 days.

Expression (53):

Days from spring tide = 6 − 3 + 2 days = 5 days.

Admiralty Tide Tables, Part II:

Table II (*a*), phase inequality of time = + 14 m.

Table II (*b*), factor for phase inequality of height = − 0·6.

Grimsby, No. 206, long. 0° 04′ W., therefore no correction for longitude to time of moon's transit.

H.W.F. & C. 5 h. 43 m., L.W.F. & C. 12 h. 11 m.

M.H.W.S. 20·1 ft., M.H.W.N. 15·6 ft., M.L. 10·7 ft.

§ 42, Rule (*a*)

Time of H.W. = 9 h. 25 m. + 5 h. 43 m. + 14 m. = 15 h. 22 m.

Rule (*b*)

Time of L.W. = 9 h. 25 m. + 12 h. 11 m. + 14 m. = 21 h. 50 m.

Rule (*c*)

Height of H.W. = $\frac{1}{2}$ (20·1 ft. + 15·6 ft.) − $\frac{1}{2}$ (20·1 ft. − 15·6 ft.) 0·6 = 16·5 ft.

Rule (*d*)

Height of L.W. = 21·4 ft. − $\frac{1}{2}$ (20·1 ft. + 15·6 ft.)
$$+ \tfrac{1}{2} \,(20\cdot1 \text{ ft.} - 15\cdot6 \text{ ft.})\, 0\cdot6 = 4\cdot9 \text{ ft.}$$

or, as the L.W. follows the H.W.:

Rule (*e*)

Time of L.W. = 15 h. 22 m. + (12 h. 11 m. − 5 h. 43 m.) = 21 h. 50 m.

Rule (*f*)

Height of L.W. = 21·4 ft. − 16·5 ft. = 4·9 ft.

Note. As Grimsby is a Secondary port the times and heights of high and low water should not have been calculated by means of the tidal constants (see § 44); the constants have, however, been used in this example in order that results may be compared with those obtained in *Example IV* (page 36). Assuming the latter to be correct, comparison will show, there being no difference between local and standard time, errors resulting from the use of the constants in this particular case; but it must not be supposed that the same errors will be found at other places or at Grimsby on other days.

§ 44. Heights of high and low water obtained by means of the tidal constants depend on the assumption that the diurnal, anomalistic and declination inequalities do not exist; times on this and on the further assumption that the ratio of $s : m$ is always 1 : 2·73.

As the neglected inequalities exist, to a greater or less extent, at all places, and as the ratio of $s : m$ varies, even in European waters, from about 1 : 7·8 at Wilhelmshaven to about 1:2·0 at Gibraltar, the times and heights found may be very considerably in error, and this method should therefore not be used when a more exact method (see Chapter IV) is available. As the ratios between the neglected inequalities and the solar and lunar semi-diurnal partial waves, and between s and m, vary at all places and vary at a particular place according to astronomical conditions, it is quite impossible to say what degree of accuracy has been attained in a particular case.

§ 45. It is possible that, were the semi-ranges of the lunar and solar semi-diurnal partial waves calculated for the particular day on which the times and heights of high and low water were required, results would reach a higher degree of accuracy, for the anomalistic and declinational inequalities do, to some extent at least, agree with theory; as, however, the diurnal partial waves may differ very considerably from their theoretical values and no information as to their intervals is available, the diurnal inequality must, in any case, be omitted.

The mean semi-ranges of the semi-diurnal partial waves may be obtained from the heights of M.H.W.S., M.H.W.N. and M.L. by means of expressions (46) and (47), and the semi-ranges on the day calculated by means of expressions (12) and (13). When these semi-ranges have been obtained, the age of the tide being known, the M.H.W.I. and M.L.W.I. may be calculated, for

$$\text{M.H.W.I.} = \text{H.W.F. \& c.} - \tfrac{1}{2} \tan^{-1} \frac{s \sin 2\theta'}{m + s \cos 2\theta'}, \qquad (57)$$

where θ' is time of moon's transit $0 - (0\cdot8 \times$ age of tide) hours, and the times and heights of high and low water may then be calculated by means of expressions (41) to (44), but it is doubtful whether the increase in accuracy of results, if any, would be worth the extra work of calculation.

§ 46. In certain cases, where the tide is of diurnal type, the constants given in the Admiralty Tide Tables, Part II, are for higher high water and lower low water; in these cases the methods described in this chapter must not be used (see also Chapter VII); the method of using, and limitations, of the higher high water and lower low water constants are fully described in the Tide Tables.

§ 47. *The times and heights of high and low water must never be found by means of the constants at either Standard or Secondary ports (see Chapter IV).*

CHAPTER IV

Approximate Tidal Prediction by means of Tidal Differences

§ **48.** As the times and heights of high and low water obtained by means of non-harmonic tidal constants may be seriously inaccurate, it is fortunate that a simple, yet more accurate, method of approximate prediction, by means of tidal differences, is available.

If two places are exposed to the action of the same tide wave, then, because the wave is a free wave, its speed depends only on the depth of water, which is invariable, and the difference between the times of arrival of either the crest or the trough at the two places should be constant; the change in range between the two places should also be constant. If one of the places is a Standard port, then the constant tidal differences for the second place may be calculated from a very few observations of the times and heights of corresponding high and low waters at the two places, and these differences applied to the predicted times and heights at the Standard port should give the corresponding times and heights at the Secondary port.

In practice it is usually found that, as the wave is compound and does not travel as a single wave, owing to obstructions and irregularities in the coastline and sea bottom between the two places and to the varying depth of water at different states of the tide, the time differences are not constant; as the ratios of $s : m$ and of the diurnal and other inequalities to the semi-diurnal partial waves are also changed as described in § 26, the height differences are not constant either; constant differences then become impossible and variable differences must be used.

It is not necessary, in the use of tidal differences, either constant or variable, that the two places should be exposed to the same tide wave; so long as the age and character of the tide, i.e. the ratio of $s : m$ and of the diurnal and other inequalities to the semi-diurnal partial waves, are the same, or approximately the same, at both places, essential conditions are fulfilled.

§ **49.** Were average high and low water time and height differences calculated for each hour of moon's transit the differences would be those of the semi-diurnal partial waves only, and their use would involve the assumption that the diurnal and other inequalities were the same at both places. Where the places are not far apart, where the range of the tide does not alter greatly between the places and where the diurnal and other inequalities are small in proportion to the semi-diurnal partial waves, this assumption is justifiable and variable differences, calculated for all times of moon's transit, in the times and heights of high and low water should, when applied to the exactly predicted times and heights of high and low water at the Standard port, give the corresponding times and heights at the Secondary port with a high degree of accuracy; so much so that, where the three necessary conditions are known to be fulfilled, variable differences are occasionally used for calculating the exact tidal predictions for Standard ports given in the Admiralty Tide Tables, Part I.

§ 50. Such variable differences are not given in the Admiralty Tide Tables, Part II, for too much space would be occupied and the necessary observations have, as a rule, not been obtained; the differences given are therefore restricted to high and low water at springs and neaps, and their use involves the assumption that changes from springs to neaps and neaps to springs are regular.

It is not always possible to calculate tidal differences, for the variations in the character of the tide are infinite and there may be no Standard port with tides of suitable character; further, though the tides at two ports may be exactly similar under certain astronomical conditions, under other conditions they may differ so materially that differences would be valueless.

At many places where differences could be calculated the observations available are insufficient; at such places the non-harmonic constants (see Chapter III) must be used; at other places the observations are insufficient for the calculation of complete spring and neap differences, at these places the following rules should be observed.

I. TIME DIFFERENCES.

(a) If a difference is bracketed under high water springs and neaps it is an average high water difference and should be used as a constant high water difference; similarly if a difference is bracketed under low water springs and neaps it should be used as a constant low water difference.

(b) If a difference is bracketed under high and low water springs and neaps it is an average difference and should be used as a constant difference for high and low water.

(c) If high water differences only are given, the high water spring difference should be used for low water springs, the high water neap difference for low water neaps. If an average high water difference, or a high water spring difference, only is given it should be used as a constant difference for high and low water.

II. HEIGHT DIFFERENCES.

(d) If a difference is bracketed under high water springs and neaps it is an average high water difference and should be used as a constant high water difference; similarly if a difference is bracketed under low water springs and neaps it should be used as a constant low water difference.

(*e*) If high water spring and neap differences only are given, low water differences should be assumed as follows:

L.W. spring difference = 0·0 ft. (58)

L.W. neap difference = H.W. spring difference
− H.W. neap difference. (59)

(*f*) If an average high water difference, or a high water spring difference, only is given, low water differences cannot be assumed and the height of low water should be calculated by § 42, rule (*f*):

To find the height of low water:

From twice the height of mean level (at the Secondary port) subtract the height of high water.

The accuracy of results obtained by means of the differences will decrease progressively from complete spring and neap differences to (*c*) and (*f*).

The assumed differences are not given in the Admiralty Tide Tables, for the rule is to give no information which does not depend on observation and which is not known to be at least approximately accurate.

§ 51. A ratio of ranges is also usually given; as this is a ratio of mean ranges, and therefore of the lunar semi-diurnal partial waves only, its use depends on the assumption that all inequalities bear the same proportion to m at the Standard and Secondary ports. At some places the height differences, at others the ratio of ranges, will give the more accurate results, but it is not possible to give any definite rule; in general, however, it is advisable to use the differences when the ratio is between 0·50 and 2·00 and to use the ratio when it is less than 0·50 or greater than 2·00.

§ 52. The use of the differences is extremely simple; when average differences are given they are applied direct to the predicted times and heights at the Standard port; when spring and neap differences are given the exact quantity to be applied in each case should be found as follows:

Difference to be applied = Spring difference
$$- \frac{(\text{Spring difference} - \text{Neap difference}) \text{ Days from spring tide}}{7}. \quad (60)$$

Days from spring tide being found by expression (53):

Days from spring tide

= Days from preceding new or full moon − age of tide,

or   Days from following new or full moon + age of tide.

The age of the tide is given in Admiralty Tide Tables, Part II, Table I, the days of new and full moon in Admiralty Tide Tables, Part I, Table IV (*b*), in the Nautical Almanac, or in the latest Supplement to Part II.

Where the ratio of ranges is to be used, the heights of high and low water should be found as follows:

Height of H.W. = M.L. at Secondary port

  + Ratio (Height of H.W. at Standard port − M.L. at Standard port). (61)

Height of L.W. = M.L. at Secondary port

  − Ratio (M.L. at Standard port − Height of L.W. at Standard port). (62)

The mean level at both the Standard and the Secondary ports must be obtained from Part II of the Tide Tables, for the difference between average mean level and mean level of the day affects both the Standard and Secondary ports.

The times of high and low water at the Secondary port, found by means of the tidal differences, will always be in the time kept at the Secondary port as given on the odd numbered pages of the Admiralty Tide Tables, Part II.

The heights of high and low water at the Secondary port, found by means of the tidal differences, will be above chart datum at the Secondary port when low water differences are given, and above a datum corresponding, allowing for difference in range, to chart datum at the Standard port when low water differences are not given.

The heights of high and low water at the Secondary port found by means of the ratio of ranges will, if the height of mean level at the Secondary port is given, be above chart datum at the port. Should a ratio be given for Secondary ports where the height of mean level is not known it must be used as a ratio of rise, that is to say, the heights of high and low water at the Standard port must be multiplied by the ratio, the results being the heights at the Secondary port above a datum corresponding, allowing for difference in range, to chart datum at the Standard port.

§ 53. *Example III.* What low water tidal differences should be assumed at Bridlington?

  Admiralty Tide Tables, Part II, No. 215:

  Bridlington, tidal differences on Immingham:

  Times, H.W.S. − 1 h. 15 m., H.W.N. − 1 h. 06 m.

  Heights, H.W.S. − 3·6 ft.,   H.W.N. − 2·9 ft.

  § 50, Rule (c), low water time differences:

  L.W.S. − 1 h. 15 m., L.W.N. − 1 h. 06 m.

Rule (e), expression (58), L.W.S. height difference = 0·0 ft.

  expression (59), L.W.N. height difference = − 3·6 ft. + 2·9 ft.

  = − 0·7 ft.

Example IV. Find the times and heights of the afternoon high and low waters at Grimsby on 3rd June, 1921, the heights being found both by means of the differences and the ratio of ranges.

  Admiralty Tide Tables, Part I:

  Table IV (b), new moon, 6th June, 1921.

Admiralty Tide Tables, Part II:

Grimsby, No. 206, tidal differences on Immingham:

Times,   H.W.S. − 0 h. 12 m.   H.W.N. − 0 h. 04 m.

   L.W.S. − 0 h. 12 m.   L.W.N. − 0 h. 03 m.

Heights, H.W.S. − 1·1 ft.   H.W.N. − 0·9 ft.

   L.W.S. − 0·1 ft.   L.W.N. − 0·3 ft.

Ratio of ranges 0·95, M.L. Grimsby 10·7 ft.

Immingham, No. 207, M.L. 11·3 ft.

Table I, age of tide, 2 days.

Expression (53):

Days from spring tide = 6 − 3 + 2 days = 5 days.

Expression (60), differences on 3rd June, 1921:

Times,   H.W., − 0 h. 12 m. − $\frac{5}{7}$ (− 0 h. 12 m. + 0 h. 04 m.) = − 0 h. 06 m.

   L.W., − 0 h. 12 m. − $\frac{5}{7}$ (− 0 h. 12 m. + 0 h. 03 m.) = − 0 h. 05 m.

Heights, H.W., − 1·1 ft.   − $\frac{5}{7}$ (− 1·1 ft. + 0·9 ft.)   = − 1·0 ft.

   L.W., − 0·1 ft.   − $\frac{5}{7}$ (− 0·1 ft. + 0·3 ft.)   = − 0·2 ft.

Admiralty Tide Tables, Part I, Immingham, 3rd June, 1921:

   H.W. time   15 h. 39 m.,   height   18·0 ft.

   Differences   − 0 h. 06 m.   − 1·0 ft.

Grimsby, H.W. time   15 h. 33 m.,   height   17·0 ft.

Admiralty Tide Tables, Part I, Immingham, 3rd June, 1921:

   L.W. time   22 h. 03 m.,   height   3·6 ft.

   Differences   − 0 h. 05 m.   − 0·2 ft.

Grimsby, L.W. time   21 h. 58 m.,   height   3·4 ft.

Using ratio of ranges, expressions (61) and (62):

Height of H.W. = 10·7 ft. + 0·95 (18·0 ft. − 11·3 ft.) = 17·1 ft.

Height of L.W. = 10·7 ft. − 0·95 (11·3 ft. − 3·6 ft.) = 3·4 ft.

Note. Heights obtained by means of the ratio of ranges are not always in exact agreement with those obtained by means of the differences (see § 51).

The above results may, as complete high and low water spring and neap differences are given, be assumed to be correct and comparison with the results obtained in Example II will show, as there is no difference between local and standard time, errors resulting from the use of the tidal constants in this particular case; but it must not be supposed that the same errors will be found at other places or at Grimsby on other days.

§ 54. In certain cases the tidal differences given in the Admiralty Tide Tables are for higher and lower high and low waters instead of for springs and neaps; in these cases the differences must be applied only to the tides to which they refer, the directions given in the tables being exactly followed.

CHAPTER V

Finding the Height of the Tide at Times between High and Low Water

§ 55. In perfect tide tables the heights of the tide would be predicted at all places and for all times; consideration of space, and the time required for the calculations, will not admit of this ideal and therefore, in the Admiralty Tide Tables, the times and heights of high and low water are predicted in Part I for certain selected Standard ports, and constants and differences, by means of which the times and heights of high and low water may be approximately calculated, are given for other ports in Part II; the times and heights of high and low water thus being known, the heights at other times may be found by so-called non-harmonic methods provided that the compound tide wave is assumed to be of harmonic form.

A harmonic wave, see § 31, is represented by the formula:

$$y = \tfrac{1}{2}r \cos h \quad \text{or} \quad -\tfrac{1}{2}r \cos h',$$

where y is the difference between sea level and mean sea level, r is the range of the wave, h the interval, or proportion of the half period of the wave expressed in degrees, between the time of high water and the time at which the height is required, and h' the corresponding proportion of the half period between the time of low water and the time at which the height is required. Then, where A is the height of mean level above chart datum, the height of the tide above chart datum is given by the expression

$$\text{Height} = A + \tfrac{1}{2}r \cos h = A - \tfrac{1}{2}r \cos h'. \tag{63}$$

§ 56. When finding the height of the tide at Standard or Secondary ports:

$$A = \tfrac{1}{2}(\text{Height of h.w.} + \text{height of l.w.}), \tag{64}$$

the heights of high and low water being those between which the required height lies; when finding the height at other ports the height of mean level as given in Part II of the Admiralty Tide Tables should be used, or, if no height of mean level is given, half the height of high water springs above chart datum.

Mean level at Standard and Secondary ports must be calculated, for this level is not constant and the mean level required is that of the tide dealt with; mean level cannot be calculated at other ports and must be assumed to be unchanging.

At all ports:

$$r = \text{Height of H.W.} - \text{height of L.W.}, \tag{65}$$

$$h = 180° \times \frac{\text{Interval from H.W.}}{\text{Duration of rise or fall}}, \tag{66}$$

$$h' = 180° \times \frac{\text{Interval from L.W.}}{\text{Duration of rise or fall}}, \tag{67}$$

the duration of rise being used when the interval is after low water or before high water, the duration of fall when the interval is after high water or before low water, and

$$\text{Duration of rise} = \text{Time of H.W.} - \text{time of L.W.}, \tag{68}$$

$$\text{Duration of fall} = \text{Time of L.W.} - \text{time of H.W.} \tag{69}$$

§ 57. The calculations required are simple, but may be shortened by tabulation of h (called θ in the Admiralty Tide Tables) for different durations of rise and fall and of $\frac{1}{2}r \cos h$ for different values of r and h. These quantities have been calculated and are given in the Admiralty Tide Tables, Part I, Tables I (*a*) and (*b*), and in Part II, Tables III (*a*) and (*b*).

Example V. Using the times and heights of high and low water at Grimsby on 3rd June, 1921 found in Example IV, find the height of the tide at 20 h. 00 m.

Example IV, H.W. 15 h. 33 m., 17·0 ft.   L.W. 21 h. 58 m., 3·4 ft.

Expression (64), $A = \frac{1}{2}$ (17·0 ft. + 3·4 ft.) = 10·2 ft.

Expression (65), r = (17·0 ft. − 3·4 ft.) = 13·6 ft.

Interval before L.W. = 21 h. 58 m. − 20 h. 00 m. = 1 h. 58 m.

Expression (69), Duration of fall = 21 h. 58 m. − 15 h. 33 m. = 6 h. 25 m.

Admiralty Tide Tables, Part I, Table I (*a*) or Part II, Table III (*a*),

$$h' = 55°.$$

Admiralty Tide Tables, Part I, Table I (*b*) or Part II, Table III (*b*),

$$\tfrac{1}{2}r \cos h' = 3·9 \text{ ft.}$$

Expression (63), Height at 20 h. 00 m. = 10·2 ft. − 3·9 ft. = 6·3 ft.

§ 58. It is sometimes necessary to find, not the height of the tide at a certain time, but the time at which the tide will reach a certain height; h then becomes the unknown, and the tables may still be used.

Example VI. Using the times and heights of high and low water at Grimsby on 3rd June, 1921 found in Example IV, what is the latest time at which a vessel drawing 17 ft. of water can enter the Royal dock, the lock sill being 6 ft. below chart datum and a clearance of 1 ft. being required between the vessel's keel and the sill?

Least necessary height of tide = 17 ft. + 1 ft. − 6ft. = 12 ft.

Example IV, H.W. 15 h. 33 m., 17·0 ft.   L.W. 21 h. 58 m., 3·4 ft.

Expression (64), $A = \frac{1}{2}(17\cdot0$ ft. $+ 3\cdot4$ ft.$) = 10\cdot2$ ft.

Expression (65), $r = (17\cdot0$ ft. $- 3\cdot4$ ft.$) = 13\cdot6$ ft.

Expression (63), $12\cdot0$ ft. $= 10\cdot2$ ft. $+ \frac{1}{2}r \cos h$.

Therefore, $\frac{1}{2}r \cos h = 1\cdot8$ ft. and interval is after H.W.

Expression (69), Duration of fall $= 21$ h. 58 m. $- 15$ h. 33 m. $= 6$ h. 25 m.

Admiralty Tide Tables, Part I, Table I (b) or Part II, Table III (b),
$$h = 75°.$$

Admiralty Tide Tables, Part I, Table I (a) or Part II, Table III (a),
Interval $= 2$ h. 40 m.

Therefore, height of tide is 12 ft. at 15 h. 33 m. $+ 2$ h. 40 m. $= 18$ h. 13 m., and this is the latest time on 3rd June, 1921, at which the vessel can enter the dock.

§ 59. The accuracy of the results obtained from the tables will depend on the accuracy of the times and heights of high and low water used and on the shape of the tide wave. A harmonic wave occupies the same time in rising and in falling and it therefore follows that, at a Standard port where the duration of rise is about equal to the duration of fall, results will probably reach a fair degree of accuracy, whereas at a port which is neither Standard nor Secondary and the duration of rise differs considerably from the duration of fall results may be seriously in error; accuracy will decrease progressively between these two extremes.

§ 60. It was shown in § 26 (d) how the type of tide may be profoundly modified when the wave is divided by land and the parts re-unite after passing the land; where the meeting waves are travelling in the same, or approximately the same, direction the tide of modified type will affect a very large area, but where directions differ greatly, and particularly where the waves are travelling in approximately opposite directions, great changes may occur within a comparatively small area, for where their crests coincide there will be a tide of great height, where the crest of one wave coincides with the trough of the other there will be little or no tide and between these extremes there will be an area of double half day tides, so called because two high and two low waters occur in each half lunar day.

§ 61. The North Sea is affected by two tide waves or rather by two portions of the Atlantic ocean tide wave which has been divided by the British islands; one portion enters through the Straits of Dover and, travelling in a north-easterly direction, causes tides on the French and Belgian North sea coasts; owing to the small width of the Straits of Dover the front of this wave is narrow, and, though it opens out after passing the Straits, it does not reach the English coast. The second portion enters on a wide front between Scotland and Norway and causes the tide in the greater part of the North sea.

The crest of one wave meets the trough of the other in the southern part of the North sea and an area, the centre of which is in about latitude 52° 45′ N., longitude 3° 20′ E., accordingly exists in which there is no appreciable tide; this area is surrounded by a region of double tides and both to northward and to southward of the area of double tides the range is increased by the partial coincidence of the crests of the waves. Double tides occur on part of the Netherlands coast but do not, as a rule, reach the English coast, though they are not unknown off Harwich and in the Thames estuary, and the small tides on parts of the coast of the Netherlands and between Harwich and Yarmouth are undoubtedly due to the meeting of the waves; north of Yarmouth and south of Harwich the range increases rapidly.

§ 62. Double half day tides also occur on the south coast of England between Southampton and Portland, their cause has not been investigated but the probability is that they are due to the tide wave, in its easterly progress up Channel, meeting a wave reflected from the French coast in the neighbourhood of the Channel islands and travelling in a northerly direction. The double tide is very different in character at different points on the coast; at Portsmouth there is no visible double high water, though the effect is seen in the slow fall of the tide for the first two hours after high water, during which only about 0·1 of the total fall occurs, whereas 0·6 of the total fall occurs in the last two hours before low water; the double low water is, however, distinctly visible between 2½ and 3 hours after the first low water, when the tide has risen about 0·4 of the total rise and, for a time, the rise ceases. At Southampton two high waters of about equal height occur, the interval between them being about 2 hours, and two low waters, as at Portsmouth of very unequal height, the second and higher about 3 hours after the first. Along the coast to the westward the interval between the double high waters increases, the double high water becomes less, and the double low water more, marked, till at Portland the predominating feature is the double low water, locally known as the *Gulder*. West of Portland the double tide is not apparent.

§ 63. For the full investigation of double tides observations at sea are required; these are not easily obtained and no complete investigations have been made. The method already described of finding the height of the tide at times between high and low water is evidently of no value where double tides occur, unless the times and heights of both high waters and both low waters are known; no alternative method is available except inside the Isle of Wight, where, owing to shallow water and the large amount of traffic, the tides are of great importance, and have been specially investigated.

§ 64. If it be assumed that the shape, not the range, of the tide wave and the duration of rise and fall at a particular place are unchanging, and if the height of the tide be observed for a complete period, from high water through low water to high water, then, where A is the height of mean level, r the range and $A + p$ the height l hours after low water on the day of the observations, on another day, when A' is the height of mean level and r' the range of the tide, l hours after low water

$$\text{Height of tide} = A' + p\,\frac{r'}{r}. \tag{70}$$

If it is known that, though the shape of the wave does not vary, the duration of rise and fall do vary, and if D be the duration of rise on the day of observation and D' the duration on the other day, then, on the other day, l hours after low water

$$\text{Height of tide} = A' + \frac{r'}{r}\left(\text{Height above mean level on}\right.$$
$$\left.\text{day of observation } l\,\frac{D}{D'} \text{ hours after low water}\right). \tag{71}$$

§ 65. In the Admiralty Tide Tables, Portsmouth is a Standard port, the times and heights of high and low water, range of tide and height of mean level are therefore known; the heights at intervals from low water have been observed and are tabulated in Part I, Table II (b), the quantity $\dfrac{D}{D'}$ has been tabulated as a correction to l in Part I, Table II (a) and the approximate height of the tide at any time may therefore be found.

Table II (a) gives a correction to be applied to the interval from low water, according to the duration of rise or fall on the day on which the height is required; applying this correction to the interval l gives the corrected interval $l\dfrac{D}{D'}$.

Table II (b) gives the quantity $\dfrac{p}{r}$ in the form of a factor.

To use the tables: after finding the range of the tide, height of mean level, duration of rise or fall and interval from low water, enter Table II (a) with the interval and duration of rise or fall and take out the corresponding correction. Correct the interval, enter Table II (b) with the corrected interval and take out the corresponding factor; multiply the range by the factor and apply the result, according to sign, to the height of mean level, the sum being the height required.

§ 66. *Example VII.* Find the height of the tide at Portsmouth at 10 h. 00 m. on 23rd February, 1921.

Admiralty Tide Tables, Part I, Portsmouth, 23rd February, 1921.
    L.W. 5 h. 00 m., − 0·5 ft. H.W. 12 h. 02 m., 13·9 ft.
    Interval after L.W. = 10 h. 00 m. − 5 h. 00 m. = 5 h. 00 m.
Expression (68), Duration of rise = 12 h. 02 m. − 5 h. 00 m. = 7 h. 02 m.
Expression (64), $A = \frac{1}{2}$ (13·9 ft. − 0·5 ft.) = 6·7 ft.
Expression (65), r = (13·9 ft. + 0·5 ft.) = 14·4 ft.

Admiralty Tide Tables, Part I, Table II (a),
    Corrected Interval = 5 h. 00 m. + 10 m. = 5 h. 10 m.
Admiralty Tide Tables, Part I, Table II (b), Factor = + 0·28.
Expression (71), Height of tide at 10 h. 00 m.
$$= 6·7 \text{ ft.} + (0·28 \times 14·4 \text{ ft.}) = 10·7 \text{ ft.}$$

§ 67. If it be assumed that, at places in the neighbourhood of Portsmouth:
  (a) Time differences on Portsmouth are constant,
  (b) Differences in mean level, referred to chart datum at the place and at Portsmouth, are constant,
  (c) Ratio of tidal movement at the place and at Portsmouth is constant,
and if the tides at such places are observed and the factor $\dfrac{p}{r}$ calculated, where p is height of tide above or below mean level at the place and r is the range at Portsmouth, then approximate heights of tide at such places may be calculated in the same manner as the Portsmouth heights, the only change being that a correction for difference in mean level, if any, must be applied to the result.

The factors at seven places in the neighbourhood of Portsmouth have been so calculated, and are tabulated in the Admiralty Tide Tables, Part I, Table II (b), and heights at these places can therefore be calculated, but it must be remembered that the interval to be used is the corrected interval from low water at Portsmouth and it is the range at Portsmouth which must be multiplied by the factor.

Corrections to mean level, where appreciable, are given at the foot of Table II (b).

§ 68. *Example VIII.* Find the height of the tide at Hurst Castle at 10 h. 00 m. on 23rd February, 1921.
Admiralty Tide Tables, Part I, Portsmouth, 23rd February, 1921,
    L.W. 5 h. 00 m., − 0·5 ft.   H.W. 12 h. 02 m., 13·9 ft.
    Interval after L.W. at Portsmouth = 10 h. 00 m. − 5 h. 00 m. = 5 h. 00 m.
Expression (68), Duration of rise at Portsmouth
$$= 12 \text{ h. } 02 \text{ m.} − 5 \text{ h. } 00 \text{ m.} = 7 \text{ h. } 02 \text{ m.}$$
Expression (64), A at Portsmouth = $\frac{1}{2}$ (13·9 ft. − 0·5 ft.) = 6·7 ft.
Expression (65), r at Portsmouth = (13·9 ft. + 0·5 ft.) = 14·4 ft.
Admiralty Tide Tables, Part I, Table II (a),
    Corrected interval after L.W. at Portsmouth
$$= 5 \text{ h. } 00 \text{ m.} + 10 \text{ m.} = 5 \text{ h. } 10 \text{ m}$$
Admiralty Tide Tables, Part I, Table II (b),
    Factor for Hurst Castle = + 0·23.
    Correction for Hurst Castle M.L. = − 3·1 ft.
Therefore, Height of tide at Hurst Castle at 10 h. 00 m.
$$= 6·7 \text{ ft.} − 3·1 \text{ ft.} + (0·23 \times 14·4 \text{ ft.}) = 6·9 \text{ ft.}$$

§ 69. Tables II (*a*) and (*b*) may also be used for finding the time at which the tide will reach a certain height, either when rising or falling, for as height and range are known the factor may be calculated and the corrected interval obtained from Table II (*b*); the true interval which, when applied to the time of low water at Portsmouth, will give the time required, may then be found from Table II (*a*).

§ 70. *Example IX.* Between what times during daylight on 25th January, 1921, can a vessel of 17 ft. draft remain at Lee-on-Solent pier, the charted depth at the pier being 6 ft. and it being considered that the vessel can proceed to the pier when a depth of 17 ft. is reached on the rising tide, but must leave again when a depth of 18 ft. is reached on the falling tide?

(*a*) Proceeding to pier:

Admiralty Tide Tables, Part I, Portsmouth, 25th January, 1921,
   L.W. 5 h. 20 m., − 0·3 ft.   H.W. 12 h. 15 m., 14·1 ft.

Duration of rise at Portsmouth   = 12 h. 15 m. − 5 h. 20 m. = 6 h. 55 m.

Expression (65), *r* at Portsmouth = 14·1 ft. + 0·3 ft. = 14·4 ft.

Expression (64), *A* at Portsmouth = ½ (14·1 ft. − 0·3 ft.) = 6·9 ft.

Admiralty Tide Tables, Part I, Table II (*b*), correction for Lee-on-Solent M.L. = 0·0 ft., therefore *A* at Lee-on-Solent = 6·9 ft.

Therefore, required height of tide above M.L.

$$= 17 \text{ ft.} - 6 \text{ ft.} - 6·9 \text{ ft.} = 4·1 \text{ ft.},$$

and
$$\text{Factor} = \frac{4·1}{14·4} = 0·28.$$

Admiralty Tide Tables, Part I, Table II (*b*), Lee-on-Solent, tide rising, factor + 0·28, corrected interval = 5 h. 12 m. after L.W. at Portsmouth.

Admiralty Tide Tables, Part I, Table II (*a*), duration of rise 6 h. 55 m., corrected interval 5 h. 12 m., approximate correction = 15 m. + to true interval or − to corrected interval.

Therefore, approximate true interval = 5 h. 12 m. − 15 m. = 4 h. 57 m.

Admiralty Tide Tables, Part I, Table II (*a*), duration of rise 6 h. 55 m., approximate true interval 4 h. 57 m., correction = 16 m. + to true interval or − to corrected interval.

Therefore, true interval = 5 h. 12 m. − 16 m. = 4 h. 56 m., and the vessel can proceed to the pier at 5 h. 20 m. + 4 h. 56 m. = 10 h. 16 m.

(*b*) Leaving pier:

Admiralty Tide Tables, Part I, Portsmouth, 25th January, 1921,
   H.W. 12 h. 15 m., 14·1 ft.   L.W. 17 h. 44 m., − 1·1 ft.

Duration of fall at Portsmouth   = 17 h. 44 m. − 12 h. 15 m. = 5 h. 29 m.

Expression (65), *r* at Portsmouth = 14·1 ft. + 1·1 ft. = 15·2 ft.

Expression (64), *A* at Portsmouth = ½ (14·1 ft. − 1·1 ft.) = 6·5 ft.

Admiralty Tide Tables, Part I, Table II (*b*), correction for Lee-on-Solent M.L. = 0·0 ft., therefore A at Lee-on-Solent = 6·5 ft.

Therefore, required height of tide above M.L.

$$= 18 \text{ ft.} - 6 \text{ ft.} - 6·5 \text{ ft.} = 5·5 \text{ ft.},$$

and
$$\text{Factor} = \frac{5·5}{15·2} = 0·36.$$

Admiralty Tide Tables, Part I, Table II (*b*), Lee-on-Solent, tide rising, factor + 0·36, corrected interval = 3 h. 04 m. before L.W. at Portsmouth.

Admiralty Tide Tables, Part I, Table II (*a*), duration of fall 5 h. 29 m., corrected interval 3 h. 04 m., approximate correction = 08 m. − to true interval or + to corrected interval.

Therefore, approximate true interval = 3 h. 04 m. + 08 m. = 3 h. 12 m.

Admiralty Tide Tables, Part I, Table II (*a*), duration of fall 5 h. 29 m., approximate true interval 3 h. 12 m., correction = 09 m. − to true interval or + to corrected interval.

Therefore, true interval = 3 h. 04 m. + 09 m. = 3 h. 13 m., and the vessel must leave the pier at 17 h. 44 m. − 3 h. 13 m. = 14 h. 31 m., and:

A vessel of 17 ft. draft can remain at Lee-on-Solent pier from 10 h. 16 m. till 14 h. 31 m. on 25th January, 1921.

Note. The difference between the corrections to the true and corrected intervals is always small; in practice therefore the correction need only be found once, instead of twice as in the example.

§ 71. Tables II (*a*) and (*b*) depend on observations obtained during one spring tide only and the methods of using the tables depend on assumptions which are not always correct in practice; results obtained must therefore be used with caution, especially at times between about 2 hours before low water and low water, when the tide falls very rapidly.

CHAPTER VI

Exact Tidal Predictions

§ 72. In the Admiralty Tide Tables, Part I, the times and heights of high and low water at a number of Standard ports in different parts of the world are predicted for each day in the year. To reach their maximum value tidal predictions should give the heights at all times, not at high and low water only, but, as the space occupied by such predictions and the time required for their calculation would be excessive, it is considered sufficient if the times and heights of high and low water only are given, leaving the heights at other times to be calculated by the approximate methods described in Chapter V.

§ 73. The predictions published in the Admiralty Tide Tables are intended to give the times and heights of high and low water of the tide due to astronomical causes as affected by terrestrial conditions and by regular meteorological changes, such as periodical variations in barometric height, rainfall, direction and force of wind, etc., but the effects of unpredictable meteorological conditions, such as gales, cannot be included; small differences are therefore usually found between the predicted and observed tides.

The height of the tide is usually increased by an abnormally low barometer and by on-shore winds, decreased by an abnormally high barometer and by off-shore winds; the effect of temporary meteorological changes, however, varies greatly at different places and no fixed rule can be given, but under normal conditions the predicted times and heights should agree with observations to within about 10 minutes and 0·5 foot respectively.

At certain places, notably in shallow estuaries, wind may affect the height of the tide very considerably, caution is therefore required when navigating such places, particularly at or near the time of low water when the barometer is high or the wind off-shore, for the depths existing may then be less than those expected.

§ 74. The necessity for tidal predictions has been felt since very early times; rough predictions for London calculated early in the thirteenth century are preserved in the British Museum; Flamsteed published more accurate predictions for the year 1683 in the *Philosophical Transactions*, vol. xiii, and it is evident that, as many harbours were not accessible at low water even to the small vessels then in use, knowledge of the times of high water was of primary importance.

Prior to the nineteenth century the methods by which predictions were calculated were secret; Whewell, in the *Philosophy of the Inductive Sciences*, remarks: "Liverpool, London and other places had their tide tables constructed by undivulged methods, which methods, in some instances at least, were handed down from father to son for several generations as a family possession, and the publication of new tables, accompanied by a statement as to their mode of calculation, was resented as an infringement of the rights of property."

The ending of this undesirable secrecy was principally due to the work of Lubbock and Whewell, which, with that of Airy, Kelvin, Darwin and others, forms the foundation of modern knowledge of the tides.

§ 75. The tides may be predicted by two methods, both of which depend on the analysis of observations, (a) non-harmonic and (b) harmonic.

(a) Regarding the tide wave as a single wave, the average lunitidal intervals and heights of high and low water, and the effect on the average intervals and heights of changes in the relative positions of the earth, moon and sun, may be calculated from the observed times and heights of high and low water; prediction by this method, when the average intervals and heights and the corrections have been calculated, consists of the correction of the average intervals and heights according to the time of transit of the moon and the declination and parallax of the sun and moon.

(b) Regarding the tide wave as a compound wave built up of a number of partial waves, the intervals and heights of the partial waves may be calculated from tidal observations of the whole compound wave, i.e. from heights of tide observed at short, regular intervals, and prediction by this method, when the intervals and heights of the partial waves have been calculated, consists of finding the height of each partial wave at the time at which the height of the tide is required, the height of the tide being the sum of the heights of the partial waves above or below mean sea level.

Neither method is exact in all cases, but, given suitable conditions, a high degree of accuracy is attained by either. Method (a), known as the Equation method because the corrections are found by a system of equations, is the more accurate, and should be used when, owing to shallow water, the partial waves are no longer of harmonic form, but accuracy decreases as the inequalities become large in proportion to the lunar semi-diurnal partial wave. Method (b), known as the Harmonic method, is the more accurate at ports so situated that the oceanic tide wave reaches them without change in shape, i.e. where all partial waves are of harmonic form, but must always be used when the inequalities, and particularly the diurnal inequality, are large in proportion to the lunar semi-diurnal partial wave.

§ 76. If the times and heights of high and low water be observed for an infinite period, then, the lunitidal intervals of all high and low waters having been calculated, the average high and low water lunitidal intervals and heights, as obtained from all the observations, will give the intervals and heights of high and low water of the lunar semi-diurnal partial wave.

If, instead of calculating averages from all the observations, these are divided into twelve groups, group 1 being intervals and heights following, allowing for the age of the tide, moon's transits 0 hr. to 1 hr. and 12 hrs. to 13 hrs., group 2 intervals and heights following transits 1 hr. to 2 hrs. and 13 hrs. to 14 hrs., ..., group 12 intervals and heights following transits 11 hrs. to 12 hrs. and 23 hrs. to 0 hr., and if the times of transit, intervals and heights in each group be meaned, high and low water lunitidal intervals and heights corresponding to twelve times of moon's transit will be obtained, and if these intervals and heights be plotted as ordinates of curves of which the times of moon's transit are the abscissae, the intervals and heights corresponding to all times of moon's transit may be read off and tabulated; these will be the intervals and heights of high and low water of the lunar semi-diurnal partial wave as affected by the phase inequality.

If now the intervals and heights in each group be divided into sub-groups according to the moon's and sun's declination and parallax, the differences between the average high and low water intervals and heights in the group and in the sub-groups will give the anomalistic, declinational and diurnal inequalities according to the time of moon's transit and moon's or sun's parallax and declination, and the tides for future times may then be predicted by summing the high and low water lunitidal intervals and heights of the lunar semi-diurnal partial wave corrected for phase inequality, and the anomalistic, declinational and diurnal inequalities, according to the time of the moon's transit and the declination and parallax of the sun and moon.

§ 77. In the method described, which is that introduced by Sir John Lubbock and described by him in the *Philosophical Transactions* for 1836, and which may be called the "method of averages," in order to ensure that all inequalities, except the one which it is desired to find, are eliminated from the averages, the observations must be continued for many years and the calculations themselves involve many months' work; as neither the observations nor the time are available, the Equation method of finding the corrections from about $1\frac{1}{4}$ years' observations by means of a system of equations, for each of which selected tides only are used, has been devised, and is described in the *Geographical Journal* for May 1919 and in the "Introductory remarks" to the Admiralty Tide Tables; this method is, however, the method of averages, described above, in an improved and abbreviated form.

§ 78. In the Harmonic method the tide wave is considered as a compound wave built up of a number of partial waves, each of constant amplitude and speed and each of harmonic form, the height of the tide at any moment, referred to mean sea level, being the sum of the heights of all the partial waves at that moment.

In this method heights are, and must always be, referred to mean sea level, and where "height" is used in the description height referred to mean sea level is intended unless otherwise stated. The heights may, if required, be referred to chart datum after the predictions have been calculated.

§ 79. In order to explain the principles of the harmonic analysis of the tide wave it is necessary, as in Chapter I, to consider separately the tidal effects of the moon and sun and of changes in their parallax and declination.

If the sun did not exist, if the earth were fixed in space and if the moon moved round the earth in the plane of the earth's equator and at a constant speed and distance, the resulting tide wave would, apart from changes due to terrestrial conditions, be a simple harmonic wave of period equal to half a lunar day, its interval and height would differ at different places but would be constant at a particular place, and from a single observed interval and height of high water the height at all future times could be predicted.

As the period of this tide is half a lunar day it may be considered as due to an imaginary satellite M revolving round the earth in half a lunar day, and

$$\text{Period of } M = \text{Half lunar day} = 12 \cdot 4 \text{ hours nearly,} \tag{72}$$

$$\text{Speed of } M = \frac{360°}{12 \cdot 4} = 29° \text{ per hour, nearly.} \tag{73}$$

§ 80. The distance of the moon from the earth is not constant, the period of the variation from perigee (nearest the earth), through apogee (farthest from the earth), to perigee being about 27 days. The moon's tide-raising force is greater than the average at perigee and less than the average at apogee, and this change in force may be considered as due to another imaginary satellite N, acting in conjunction with M at perigee and in opposition to M at apogee.

The speed of M is twice that of the moon, so M and N as imagined will act in conjunction when 0° apart and in opposition when 180° apart, not, as is the case with the true sun and moon, in conjunction when 0° and 180° apart and in opposition when 90° apart.

M revolves round the earth in 12·4 hours; in $13\frac{1}{2}$ days N must lose (or gain) 180° on M, therefore

$$\text{Speed of } N = 29° - \frac{180°}{13 \cdot 5 \times 24} = 28°5 \text{ per hour nearly,} \tag{74}$$

$$\text{Period of } N = \frac{360°}{28°5} = 12 \cdot 6 \text{ hours nearly,} \tag{75}$$

and $\qquad$ Amplitude of resulting tide at perigee $= M$ tide $+ N$ tide, $\qquad$ (76)

$\qquad$ Amplitude of resulting tide at apogee $= M$ tide $- N$ tide, $\qquad$ (77)

the imaginary satellites being on the meridian simultaneously at perigee and 180° apart at apogee.

The periods and speeds of M and N being known, their times of transit may be calculated and the intervals and heights of the resulting partial waves found from tidal observations.

The time of transit of a satellite and the interval and height of the partial wave due to it having been calculated the height of the partial wave at any time may be found from the harmonic formula, see § 55, $y = \frac{1}{2}r \cos h$.

§ 81. The partial waves due to the moon's movement in declination, to the mean sun and to the sun's movements in parallax and declination may be similarly considered as due to imaginary satellites, each revolving round the earth in the plane of the earth's equator and at a constant speed and distance, and the height of the tide at any time is the sum of the heights of all the partial waves, each considered as due to an imaginary satellite, at that time.

In order to obtain a true representation of the tides as they exist a considerable number of satellites must be assumed, and the computing of predictions becomes a lengthy process; computation is, however, avoided by the use of the predicting machine, first suggested by Lord Kelvin and brought into practical form by Mr Edward Roberts.

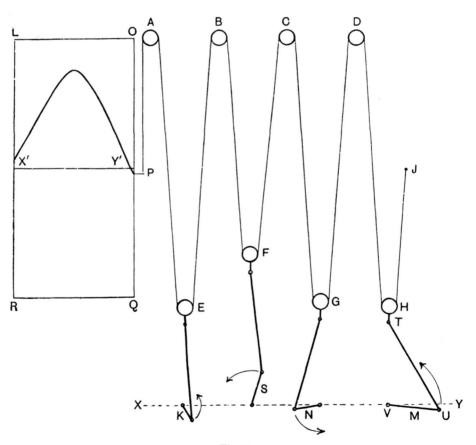

Fig. 8.

§ 82. Figure 8 represents diagrammatically a machine for predicting tides by means of four partial waves; in practice no machine which includes less than about 20 partial waves should be used for accurate predictions, and machines have been constructed to include no less than 40.

In the figure, *LOQR* is a cylindrical drum; *A*, *B*, *C*, *D* are fixed pulleys, *E*, *F*, *G* and *H* are pulleys moveable in an up and down direction; a cord fixed at *J* and terminating in a pen *P*, in contact with the drum, runs over all pulleys.

The moveable pulleys are connected to cranks *K*, *S*, *N*, *M* by means of rods, the pulley *H* being connected to the crank *M* by means of the rod *TU* pivoted at *T* and *U*, and *M* is free to revolve round the fixed point *V*. *G*, *F* and *E* are similarly connected to their cranks, which are also free to revolve.

The length of the string is adjusted so that when all cranks are on the horizontal line *XY* the pen is on the horizontal line *X'Y'*.

If the crank *M* revolves in the direction shown by the arrow, movement will be transmitted to the pen; if the drum also revolves, and if the relative rates of revolution of the drum and crank are known, the position which the crank occupied at any time will be shown by the curve traced on the drum.

If the length of *M* be adjusted so as to represent on a suitable scale the amplitude of the partial wave *M*, if one revolution of the drum be considered to represent half a mean solar day, and if *M* be constructed so as to make $\frac{24}{24 \cdot 8} = 0 \cdot 97$ revolution, nearly, whilst the drum revolves once, the curve traced will represent the partial wave *M* on a scale equal to twice the scale to which the crank has been adjusted.

If the hour angle of the satellite *M* at, say, 0 h. on 1st January 1921, be known, and if the crank be set to an angle equal to this hour angle plus the interval of the partial wave, the curve traced will, if the machine work continuously, represent the partial wave *M* for the year 1921.

As the pen sums the heights of all the cranks, if each crank is adjusted in length and set so as to correctly represent the state of a partial wave at the time when it is desired to commence the predictions, and if the rate of revolution of each crank bears the same proportion to the rate of revolution of the drum as the period of the partial wave represented bears to a half mean solar day, the curve traced will correctly represent the compound tide wave, and from it the times and heights of high and low water, or the height of the tide at any time, may be obtained.

As the periods of the partial waves are the same at all ports, the intervals and amplitudes only differing, and as the cranks are adjustable both for interval and amplitude, the machine may be used for predicting the tide of any port provided that the constants of the partial waves are known.

When the constants of the more important partial waves are known approximate heights may be computed, without the machine, as described in Chapter VII, where also a further explanation of harmonic methods is given.

CHAPTER VII

Approximate Tidal Prediction by means of Harmonic Tidal Constants

§ 83. Where, as in the waters surrounding the British islands, the tide is almost entirely semi-diurnal, the times and heights of high and low water and the height of the tide at other times, may be approximately found by the non-harmonic methods described in Chapters III and V; where, however, the diurnal partial waves are large in proportion to the semi-diurnal these methods fail and harmonic constants must be used.

Harmonic constants for the calculation of approximate predictions have not yet been adopted in the Admiralty Tide Tables or, in fact, in any official tables with the exception of those for the Netherlands East Indies, but as this is the only method, apart from tidal differences on Standard ports, by means of which such predictions can be calculated for places at which the diurnal partial waves are large in proportion to the semi-diurnal, and as, in such cases, even tidal differences may fail, its universal adoption is only a matter of time.

The principles of the harmonic method of tidal prediction and the use of the predicting machine were briefly explained in Chapter VI; exact tidal predictions require the constants of a large number of partial waves, calculated from a long series of observations, and the use of the machine, but approximate heights may be computed by means of the constants of the more important partial waves, and these may be calculated from a comparatively short series of observations.

§ 84. A universal notation has been adopted for the harmonic method and is used in the following explanation. "*Satellite*" means an imaginary satellite, and a *component* is the partial wave due to the satellite; the *speed* of a satellite or component is its angular velocity in degrees per hour; *hours, days,* and *years* refer to mean solar hours, days and years and *time* to mean solar time, unless *satellite days* or *time* or *component days* or *time* are specially mentioned.

A *satellite day* or *component day* is the period of the satellite or component and *satellite* or *component time* is measured in degrees, 360 degrees to a satellite or component day.

The *harmonic constants* of a component are H, the semi-amplitude, and k, the interval, in satellite or component time, between the transit of the satellite and high water of the component.

Each satellite or component is distinguished by a capital letter, followed by a small numeral which indicates approximately the number of times which the satellite crosses the meridian, or the component recurs, in a solar day; thus M_2 indicates a satellite which crosses the meridian twice in a day, or the component, due to the satellite, which makes high water twice in a day, O_1 is a diurnal satellite or component and M_4 crosses the meridian, and makes high water, four times in a day.

§ 85. If the sun did not exist, if the earth were fixed in space and if the moon revolved round the earth in the plane of the earth's equator and at a constant speed and distance, a simple tide wave would result, and were a single observation of the semi-amplitude (H) and interval (k) of this wave obtained the height of the tide at any future time could be predicted by means of the harmonic expression

$$y = H \cos n \{t - (T + k)\}, \tag{78}$$

where y is the difference between water level and mean water level, H is the semi-amplitude of the wave, n the speed of the wave, t the time at which the height is required, T the time of the moon's transit and k the interval between moon's transit and time of high water of the wave.

If, instead of a single moon there were a number of satellites, each revolving round the earth in the plane of the earth's equator and at a constant speed and distance, the wave resulting would be compound, and would be made up of the simple waves, or components, due to each satellite, H and k of each component could be found from observation by a system of analysis and the expression for the height of the compound wave at any time would be

$$y = H_1 \cos n_1 \{t - (T_1 + k_1)\} + H_2 \cos n_2 \{t - (T_2 + k_2)\}$$
$$+ H_3 \cos n_3 \{t - (T_3 + k_3)\} + \dots \text{etc.} \tag{79}$$

§ 86. In the harmonic method of prediction the tidal forces of the sun and moon in their variable paths are considered as due to a number of satellites, each revolving round the earth in the plane of the earth's equator and at a constant speed and distance, so chosen that the combined tidal forces of the satellites at any moment exactly equal the tidal forces of the true sun and moon at that moment; the compound tide wave at any place, due to the satellites, therefore exactly represents the tide wave at that place as it occurs in reality.

The harmonic method of prediction thus comprises:

(*a*) The division of the true tide wave, as obtained from observations, into a number of simple waves, each of the form of expression (78) and each of which may be considered as due to a satellite revolving round the earth in the plane of the earth's equator and at a constant speed and distance.

(*b*) The re-combination of the simple waves in accordance with expression (79) into the tide wave as it will exist at a future time.

As this is the only method by which the tides may be predicted, either exactly or approximately, over the large portion of the earth's surface where the diurnal partial waves are large in proportion to the semi-diurnal, it is of increasing importance; in order that computation by this method may become possible to seamen its principles must be understood and will therefore be explained in some detail, though only the satellites, or components, required for approximate predictions will be traced.

In explaining principles it is convenient to refer to satellites, but it must be remembered that each satellite generates a component having the same speed.

§ 87. The *absolute* motions of the earth and moon are as follows:

(i) The earth rotates on its axis in 23·9345 hours, this being the length of a sidereal day, or interval between successive transits of a fixed star across the meridian of a place; the speed of this motion is:

$$e = \frac{360°}{23\cdot9345} = 15°0411 \text{ per hour.} \tag{80}$$

(ii) Relatively to the fixed stars, the moon makes a complete revolution round the earth in 27·3216 days, this being the length of the tropical month, or period of the moon's declination; the speed of this motion is:

$$m = \frac{360°}{24 \times 27\cdot3216} = 0°5490 \text{ per hour.} \tag{81}$$

(iii) The earth revolves round the sun in 365·2422 days; the speed of this motion is:

$$s = \frac{360°}{24 \times 365\cdot2422} = 0°0411 \text{ per hour.} \tag{82}$$

(iv) The position of the axis of the ellipse in which the moon revolves round the earth is not constant but makes a complete revolution in 8·85 years, consequently the anomalistic month, or interval between successive passages of the moon through perigee, differs slightly from the tropical month; the speed of this motion is:

$$p = \frac{360°}{8\cdot85 \times 24 \times 365\cdot2422} = 0°0046 \text{ per hour.} \tag{83}$$

§ 88. From the absolute motions *relative* speeds may be found by simple addition or subtraction, thus:

(i) The speed of the sun with respect to a point on the earth's surface is $(e - s)°$:

$$(e - s)° = 15°0411 - 0°0411 = 15°0000 \text{ per hour.} \tag{84}$$

(ii) The length of a mean solar day is $\dfrac{360°}{(e - s)°}$ hours:

$$\frac{360°}{(e - s)°} = \frac{360°}{15°} = 24\cdot0000 \text{ hours.} \tag{85}$$

(iii) The speed of the moon with respect to a point on the earth's surface is $(e - m)^\circ$:

$$(e - m)^\circ = 15^\circ 0411. - 0^\circ 5490 = 14^\circ 4921 \text{ per hour.} \qquad (86)$$

(iv) The length of a mean lunar day is $\dfrac{360^\circ}{(e - m)^\circ}$ hours:

$$\frac{360^\circ}{(e - m)^\circ} = \frac{360^\circ}{14^\circ 4921} = 24\cdot 8412 \text{ hours.} \qquad (87)$$

(v) The speed of the moon in its orbit with respect to the axis of the ellipse is $(m - p)^\circ$:

$$(m - p)^\circ = 0^\circ 5490 - 0^\circ 0046 = 0^\circ 5444 \text{ per hour.} \qquad (88)$$

(vi) The length of the anomalistic month is $\dfrac{360^\circ}{24\,(m - p)^\circ}$ days:

$$\frac{360^\circ}{24\,(m - p)^\circ} = \frac{360^\circ}{24 \times 0^\circ 5444} = 27\cdot 5546 \text{ days.} \qquad (89)$$

(vii) The speed of the moon with respect to the sun, as seen from a point on the earth's surface, is $(m - s)^\circ$:

$$(m - s)^\circ = 0^\circ 5490 - 0^\circ 0411 = 0^\circ 5079 \text{ per hour.} \qquad (90)$$

(viii) The length of the synodic month, or interval between successive occurrences of the same phase of the moon, is $\dfrac{360^\circ}{24\,(m - s)^\circ}$ days:

$$\frac{360^\circ}{24\,(m - s)^\circ} = \frac{360^\circ}{24 \times 0^\circ 5079} = 29\cdot 5306 \text{ days.} \qquad (91)$$

§ 89. The mean moon revolves round the earth in the plane of the earth's equator at a constant speed and distance and causes a simple wave of period half a lunar day. The mean lunar tide may therefore be considered as due to a satellite M_2 revolving round the earth in the plane of the earth's equator, at a constant distance and at a constant speed of twice the average speed of the moon, and:

$$\text{Speed of } M_2 = 2\,(e - m)^\circ = 28^\circ 9842 \text{ per hour,} \qquad (92)$$

$$\text{Period of } M_2 = \frac{360^\circ}{28^\circ 9842} = 12\cdot 4206 \text{ hours.} \qquad (93)$$

§ 90. The true moon, however, does not remain in the plane of the earth's equator and, though its velocity is constant, its distance, and therefore its speed, is not.

Considering first the variation in distance and speed, the problem is to substitute for the single true moon, moving in the plane of the earth's equator at a varying distance and speed, a satellite M_2, moving in the plane of the earth's equator at the average distance of the true moon but with twice its average speed, and one or more additional satellites, each moving in the plane of the earth's equator at a constant distance and speed, in such a manner that the collective forces of M_2 and of the additional satellites shall equal the tidal force of the true moon.

Following the method in § 88, the average speed of the true moon with respect to a point on the earth's surface is $(e - m)°$ and the speed of M_2, representing the tidal force of the mean moon, is $2 (e - m)°$, whilst the speed of the true moon or of M_2 with respect to the axis of the orbit is $(m - p)°$, consequently the satellite N_2 will fulfil the required conditions if:

$$\text{Speed of } N_2 = 2 (e - m)° - (m - p)° = 28°4398 \text{ per hour,} \tag{94}$$

$$\text{Period of } N_2 = \frac{360°}{28°4398} = 12·6584 \text{ hours,} \tag{95}$$

for the difference between the speeds of M_2 and N_2

$$= 28°9842 - 28°4398 = 0°5444 \text{ per hour,}$$

and therefore if, at any time, the tidal forces of M_2 and N_2 are acting in conjunction (perigee) they will, after the lapse of half an anomalistic month, be acting in opposition (apogee), for:

$$0°5444 \times 24 \times 13·7773 = 180°, \tag{96}$$

and after the further lapse of half an anomalistic month they will again be acting in conjunction, for:

$$0°5444 \times 24 \times 27·5546 = 360° = 0°. \tag{97}$$

§ 91. Considering now changes in the moon's declination; the moon does not move in the plane of the earth's equator but in a path which, on the average, coincides with the ecliptic but which oscillates about this average, the maximum divergence being about 5° on either side and the period of the oscillation about 18·7 years. The axis of the moon's path is therefore not perpendicular to the earth's axis, the effect being that, as shown by expression (5):

(a) When the moon's declination is not zero the two tides of the day are of unequal range, the inequality increasing with the declination.

(b) The moon's tidal force decreases from a maximum when declination is zero to a minimum at maximum declination,

and, owing to the change in maximum declination,

(c) The tidal forces of the satellites M_2 and N_2 and of additional satellites, introduced to represent changes in tidal force due to changes in declination, are not constant.

The inequality (a) may be represented by assuming one or more satellites causing tidal oscillations of a diurnal character, the conditions to be fulfilled being that the tidal forces of the satellites have a period of half a tropical month, or 13·6608 days, for, as the angle between the axis of the moon's path and the plane of the earth's equator is the same for north and south declination, the effect must be the same for north and south declination, and also that the tidal forces of the satellites are maximum at extreme declination and vanish when declination vanishes.

The earth and moon occupy the same positions relatively to the fixed stars every $27 \cdot 3216$ days and the speed of this motion is $m°$; the speed of the moon with respect to a point on the earth's surface is $(e - m)°$ and the satellites K_1 and O_1 fulfil the above conditions when:

$$\text{Speed of } K_1 = (e - m)° + m° = e° = 15°0411 \text{ per hour,} \tag{98}$$

$$\text{Speed of } O_1 = (e - m)° - m° = (e - 2m)° = 13°9431 \text{ per hour,} \tag{99}$$

$$\text{Period of } K_1 = \frac{360°}{15°0411} = 23\cdot9345 \text{ hours,} \tag{100}$$

$$\text{Period of } O_1 = \frac{360°}{13°9431} = 25\cdot8193 \text{ hours.} \tag{101}$$

The joint action of these satellites fulfils the conditions, the action of either by itself does not, for if, at a given moment when declination is extreme, the forces of the satellites are acting together, they will balance after the lapse of one-quarter of a tropical month, or $6\cdot8304$ days, when declination is zero, for:

$$6\cdot8304 \times 24 \times (15°0411 - 13°9431) = 180°, \tag{102}$$

and after the lapse of half a tropical month, or $13\cdot6608$ days, when declination is again extreme, they will again be acting together, for:

$$13\cdot6608 \times 24 \times (15°0411 - 13°9431) = 360° = 0°. \tag{103}$$

Inequality (b) requires the same conditions for its fulfilment, but, as the effect is the same on both lunar tides, a single satellite K_2, of twice the speed and half the period of K_1, will meet the case, and:

$$\text{Speed of } K_2 = 2 \times 15°0411 = 30°0822 \text{ per hour,} \tag{104}$$

$$\text{Period of } K_2 = \tfrac{1}{2} \times 23\cdot9345 = 11\cdot9673 \text{ hours.} \tag{105}$$

§ 92. The tidal forces of the satellite M_2, representing the mean moon and therefore a moon in average declination, are not constant but are greatest in years of least declination, and least in years of greatest declination (§ 91 (c)); the tidal forces of N_2 vary in a similar manner; the forces of K_1, O_1 and K_2, on the other hand, are greatest in years of greatest declination and least in years of least declination. The tidal forces of all the satellites which together represent the true moon therefore require correction by means of factors which vary according to the moon's average declination of the year the tides of which are to be analysed or predicted; the factors used depend on theoretical considerations and are the same for M_2 and N_2, but as K_1 and K_2 are luni-solar satellites (see below) the factors for these and for O_1 differ.

§ 93. The satellites and component waves due to the sun may be found by similar reasoning; thus the satellite S_2, representing the mean sun, has:

$$\text{Speed} = 2s° = 30°0000 \text{ per hour,} \tag{106}$$

$$\text{Period} = \frac{360°}{30°0000} = 12\cdot0000 \text{ hours.} \tag{107}$$

The satellites K_1 and P_1, representing the diurnal effect of changes in the sun's declination, have:

$$\text{Speed of } K_1 = (e - s)° + s° = e° = 15°0411 \text{ per hour,} \tag{108}$$

$$\text{Speed of } P_1 = (e - s)° - s° = (e - 2s)° = 14°9589 \text{ per hour,} \tag{109}$$

$$\text{Period of } K_1 = \frac{360°}{15°0411} = 23\cdot9345 \text{ hours,} \tag{110}$$

$$\text{Period of } P_1 = \frac{360°}{14°9589} = 24\cdot0659 \text{ hours,} \tag{111}$$

and the satellite K_2, representing the semi-diurnal effect of changes in the sun's declination, has:

$$\text{Speed} = 2 \times 15°0411 = 30°0822 \text{ per hour,} \tag{112}$$

$$\text{Period} = \tfrac{1}{2} \times 23\cdot9345 = 11\cdot9673 \text{ hours.} \tag{113}$$

It should be noted that the speeds and periods of K_1 and K_2 of the moon and sun are the same, the lunar and solar effects of these satellites can therefore not be separated and they are considered as luni-solar satellites, the corresponding component waves being considered as luni-solar waves.

§ 94. For approximate tidal predictions it is not necessary to consider further satellites, or components, due to the tidal forces of the sun and moon, but it is evident that these exist, and are required for accurate predictions, for there must be a satellite L_2 of speed $2 (e - m)° + (m - p)°$, there must be solar satellites corresponding to N_2 and L_2 of the moon etc.; the gain in accuracy due to the inclusion of the component waves due to these satellites is, however, not sufficient to compensate for the increased difficulty of computing approximate tidal heights, but where accurate predictions are required and the machine is to be used, all must be included. Components due to terrestrial causes will now be considered.

§ 95. Waves in deep water are of harmonic form; when a harmonic wave enters shallow water it undergoes a change of shape, the front slope becoming more, and the rear less, steep. The components, or *over-tides*, introduced as corrections to the solar and lunar components in order to show this change in shape, are themselves of harmonic form and each lunar and solar component instead of being formed in accordance with expression (78) thus takes the shape of expression (79). The speeds of the over-tides of any component are

2, 3, 4, etc. times the speed of the component, their periods being $\frac{1}{2}$, $\frac{1}{3}$, $\frac{1}{4}$, etc. the period of the component, and, preserving the original notation, the over-tides of the component representing the tidal force of the mean moon are M_4, M_6, M_8, etc.

It is only necessary, for the purpose of approximate predictions, to con-sider M_4, the first over-tide of the lunar semi-diurnal component M_2; even this need only be considered at ports situated on rivers or estuaries where, owing to decreasing depth and converging shore lines, the tide wave undergoes great change in shape, and

$$\text{Speed of } M_4 = 2 \times 28°9842 = 57°9684 \text{ per hour,} \qquad (114)$$

$$\text{Period of } M_4 = \tfrac{1}{2} \times 12 \cdot 4206 = 6 \cdot 2103 \text{ hours.} \qquad (115)$$

§ 96. When two waves meet in deep water the height of the resulting wave at any moment is equal to the sum of the heights of the meeting waves at that moment; in shallow water this is not the case, but it can be shown that the result of the interaction of two harmonic waves in shallow water may be represented by introducing two additional harmonic waves of speeds equal to the sum and difference of the speeds of the interacting waves. The additional components so introduced are called *compound tides*, but for approximate predictions it is only necessary to consider MS_4, the compound of the lunar and solar semi-diurnal components M_2 and S_2 and of speed equal to the sum of their speeds; even this, as M_4, need only be considered at ports situated on rivers or estuaries, and:

$$\text{Speed of } MS_4 = 28°9842 + 30°0000 = 58°9842 \text{ per hour,} \qquad (116)$$

$$\text{Period of } MS_4 = \frac{360°}{58°9842} = 6 \cdot 1917 \text{ hours.} \qquad (117)$$

As M_4 is a lunar, and MS_4 a luni-solar, component neither is constant and both vary from year to year according to the moon's declination (see § 92), but as the components are invariably small the change may be neglected.

§ 97. The components required for the calculation of approximate tidal heights by the harmonic method may now be summarised as follows:

M_2, the principal lunar semi-diurnal component, speed 28°9842 per hour, period 12·4206 hours.

S_2, the principal solar semi-diurnal component, speed 30°0000 per hour, period 12·0000 hours.

N_2, the larger lunar elliptical semi-diurnal component, speed 28°4398 per hour, period 12·6584 hours.

K_1, the luni-solar declinational diurnal component, speed 15°0411 per hour, period 23·9345 hours.

O_1, the lunar declinational diurnal component, speed 13°9431 per hour, period 25·8193 hours.

K_2, the luni-solar declinational semi-diurnal component, speed $30°0822$ per hour, period $11·9673$ hours.

P_1, the solar declinational diurnal component, speed $14°9589$ per hour, period $24·0659$ hours.

M_4, the first over-tide of M_2, speed $57°9684$ per hour, period $6·2103$ hours.

MS_4, compound of M_2 and S_2, speed $58°9842$ per hour, period $6·1917$ hours.

M_4 and MS_4 being usually only required at ports situated on rivers or estuaries.

The speeds and periods of the components are the same for all places; in recording the harmonic constants of a port it is therefore only necessary to give H (the semi-amplitude) and k (the interval) of each component as found from tidal observations.

§ 98. The height of tide found by means of the constants is above or below mean sea level; for navigational purposes the height must be referred to chart datum and the difference between the height of mean sea level and chart datum must therefore be added to all heights obtained from the constants. This difference is regarded as an additional component, and

A_0 is the height of mean sea level above chart datum.

Where A_0 is not exactly known, and is not given with the harmonic constants, it should be assumed that

$$A_0 = H \text{ of } (M_2 + S_2 + K_1 + O_1). \tag{118}$$

§ 99. The calculation of harmonic constants is not usually required of seamen; the methods of calculating the more important constants from a short series of observations are described in the *Admiralty Manual of Scientific Enquiry* (Eyre and Spottiswoode Limited, East Harding Street, Fleet Street, London, E.C.), a more complete account of the calculation of the constants required for exact prediction being given in *A Manual of Tidal Observations*, by Major A. W. Baird, R.E., F.R.S. (Taylor and Francis, Red Lion Court, Fleet Street, London, E.C.).

Briefly the methods adopted depend on the fact that, if, calling one-twenty-fourth of the period of a component a *component hour*, the heights of the tide be observed and tabulated for each component hour of a number of component days, then the effect of all other components will be eliminated from the mean of all the heights observed at each component hour.

§ 100. The times of transit of the satellites may be calculated, and, if H and k of the components are known:

$$\text{Time of h.w. of a component} = \text{Time of transit of satellite} + \frac{k}{\text{Speed}}, \tag{119}$$

$$\text{Height of a component wave at any time} = H \cos I, \tag{120}$$

where I is the interval from h.w. multiplied by the speed of the component,

Height of tide at any time, referred to chart datum

$$= \text{Algebraic sum of heights of all components at that time and of } A_0. \tag{121}$$

§ 101. The calculations required by expressions (119) to (121) are simple but, where up to nine components have to be dealt with, occupy much time; by means of Tables I to VI, given in the Appendix, calculation is almost entirely avoided and the height of the tide at any moment at any place where the harmonic constants are known may be approximately found in a few minutes.

If the times and heights of high and low water are required the heights must be found for a series of hours and plotted as ordinates of a curve of which times form the abscissae; the times and heights of high and low water may then be read off from the curve (see Figure 9).

Table I gives the astronomical arguments of the satellites at 0 h. G.M.T. on 1st January (midnight between 31st December and 1st January) of each year from 1920 to 1950; the astronomical argument of a satellite is its component time at 0 h. at Greenwich and will, if divided by its speed, give its hour angle in solar time at 0 h. at Greenwich.

If the astronomical arguments are required for other days and other meridians, those given must be corrected both for the days elapsed since 1st January and, where the period of the satellite differs appreciably from 12 h. or a multiple of 12 h., for longitude.

Table II gives the correction for longitude for the satellites M_2, O_1, MS_4, N_2 and M_4; the astronomical arguments of S_2, K_1, K_2 and P_1 do not require correction. The correction for longitude is to be subtracted in east longitude, added in west longitude.

Table III gives the increment, or correction, for the days elapsed since 1st January; this correction is always added, and it must be particularly noted that, in leap years, one day must be added to all dates after 28th February in order to allow for the extra day in that month.

Table IV. By means of this table intervals of mean solar time may be converted into component time, according to the speeds of the satellites.

Table V gives the factors by which H of the lunar components must be multiplied in order to obtain the correct H for any year, see § 92.

Table VI gives the value of $H \cos I$, see expression (120).

As harmonic constants are not at present given in Admiralty publications, Table VII, giving the constants for certain selected ports, has been added.

§ 102. To find the height of a component at any time:

In order to avoid confusion between component time and mean solar time all operations are performed in component time, and

(*a*) From Table I, take the astronomical argument of the satellite for 0 h., on 1st January for the year.

(*b*) Correct the astronomical argument by Table II, if necessary.

(*c*) Take the increment from Table III according to date, adding one day if the year is a leap year.

(d) Add together the corrected astronomical argument, the increment and k of the component and diminish the sum by 360° if between 360° and 720° or by 720° if above 720°; the result is the local component time of high water of the component and could be converted into local mean solar time, if required, by dividing by the speed of the component.

As the period of any satellite, or component, is 360° in its own time, diminishing by 360° or 720° causes no change in time.

(e) Convert the solar time at which the height is required into component time by means of Table IV.

(f) Take the difference between the component time of high water (d) and the component time at which the height is required (e), thus obtaining I, expression (120).

(g) Correct H of the component, if necessary, by means of Table V.

(h) Enter Table VI with corrected H and I (see footnote to table, and take out H cos I, the height required.

§ 103. *Example X.* Find the height, referred to mean sea level, of the component M_2 at Shanghai (Wusung bar) at 12 h. on 25th January, 1921, longitude of Wusung 121° 30′ E.

(a) Table I, astronomical argument at 0 h. 1st January, 1921		159°
(b) Table II, correction for longitude		− 8
Corrected astronomical argument . . .		151
(c) Table III, increment for 25th January		225
Table VII, k of M_2 at Shanghai		30
(d) Sum		406
Diminish by		360
Component time of H.W.		46
(e) Table IV, 12 h. in component time		348
(f) Difference = I		302
Table VII, H of M_2 at Shanghai		3·11 ft.
(g) Table V, factor for 1921		1·03
Corrected H of M_2		3·20
(h) Table VI, H cos I = height of component at 12 h. on 25th January		+ 1·70 ft.

Note. The result obtained at (h) is the same whichever way the subtraction at (f) is made, but, as the height of the component may be required for a series of hours, it is essential to know whether it is increasing or decreasing; the component time of high water should therefore always be subtracted from the component time at which the height is required, when I will always be increasing and, 0° being high water and 180° low water of the component, the height of the component is:

(i) *Decreasing above* mean sea level when I is between   0° and 90°

(ii) *Increasing below*   ,,   ,,   ,,   90 ,, 180

(iii) *Decreasing below*   ,,   ,,   ,,   180 ,, 270

(iv) *Increasing above*   ,,   ,,   ,,   270 ,, 360

§ 104. The height of a component above chart datum is found by adding the height obtained as above to A_0, the height of mean sea level above chart datum, and the height of the tide above chart datum by adding the sum of the heights of all components to A_0.

Example XI. Find the height of the tide, above chart datum, at Shanghai (Wusung bar) at 12 h. on 25th January, 1921, longitude of Wusung 121° 30′ E.

	M_2	S_2	N_2	K_1	O_1	K_2	P_1	M_4	MS_4	A_0	
Table I	159°	0°	230°	345°	175°	152°	10°	318°	159°		
Table II	−8	0	−12	0	−8	0	0	−16	−8		
Table III ...	225	0	179	336	249	313	24	90	225		
Table VII, k ...	30	77	2	207	149	77	207	331	18		
Sum	406	77	399	888	565	542	241	723	394		
Diminish by ...	360	0	360	720	360	360	0	720	360		
H.W. 25th January	46	77	39	168	205	182	241	3	34		
Table IV, 12 h. ...	348	0	341	180	167	1	180	336	348		
I	302	283	302	12	322	179	299	333	314		
	ft.	ft.	ft.	ft.	ft.	ft.	ft.	ft.	ft.	ft.	
Table VII, H ...	3·11	1·03	0·40	0·66	0·46	0·28	0·22	0·70	0·47	6·55	
Table V, factors	1·03	—	1·03	0·89	0·86	0·77	—	—	—		
Corrected H ...	3·20	1·03	0·41	0·59	0·40	0·22	0·22	0·70	0·47	6·55	
Table VI, $H \cos I$	+1·70	+0·24	+0·22	+0·58	+0·31	−0·22	+0·11	+0·62	+0·33	+6·55	10·44 ft.

Sum = height above chart datum

§ 105. When the height of the tide is required each hour for a series of hours it is only necessary to find the component time of high water once for each component, for the hourly change in I is then given by Table IV, and to correct H of each component once, for the corrected H remains constant for a year.

Example XII. Find the height of the tide, above chart datum, at Shanghai (Wusung bar) each hour from 0 h. to 23 h. on 25th January, 1921, longitude of Wusung 121° 30′ E.

From Example XI: H.W. 25th January	M_2	S_2	N_2	K_1	O_1	K_2	P_1	M_4	MS_4	A_0
	46°	77°	39°	168°	205°	182°	241°	3°	34°	
Corrected H	3·20 ft.	1·03 ft.	0·41 ft.	0·59 ft.	0·40 ft.	0·22 ft.	0·22 ft.	0·70 ft.	0·47 ft.	6·55 ft.

Solar Time	M_2		S_2		N_2		K_1		O_1		K_2		P_1		M_4		MS_4	
	Table IV	I	Table IV	I	Table IV	I	Table IV	I	Table IV	I	Table IV	I	Table IV	I	Table IV	I	Table IV	I
h.	46°		77°		39°		168°		205°		182°		241°		3°		34°	
0	0°	314°	0°	283°	0°	321°	0°	192°	0°	155°	0°	178°	0°	119°	0°	357°	0°	326°
1	29	343	30	313	28	349	15	207	14	169	30	208	15	134	58	55	59	25
2	58	12	60	343	57	18	30	222	28	183	60	238	30	149	116	113	118	84
3	87	41	90	13	85	46	45	237	42	197	90	268	45	164	174	171	177	143
4	116	70	120	43	114	75	60	252	56	211	120	298	60	179	232	229	236	202
5	145	99	150	73	142	103	75	267	70	225	150	328	75	194	290	287	295	261
6	174	128	180	103	171	132	90	282	84	239	180	358	90	209	348	345	354	320
7	203	157	210	133	199	160	105	297	98	253	211	29	105	224	46	43	53	19
8	232	186	240	163	228	189	120	312	112	267	241	59	120	239	104	101	112	78
9	261	215	270	193	256	217	135	327	125	280	271	89	135	254	162	159	171	137
10	290	244	300	223	284	245	150	342	139	294	301	119	150	269	220	217	230	196
11	319	273	330	253	313	274	165	357	153	308	331	149	165	284	278	275	289	255
12	348	302	0	283	341	302	180	12	167	322	1	179	180	299	336	333	348	314
13	17	331	30	313	10	331	196	28	181	336	31	209	194	313	34	31	47	13
14	46	0	60	343	38	359	211	43	195	350	61	239	209	328	92	89	106	72
15	75	29	90	13	67	28	226	58	209	4	91	269	224	343	150	147	165	131
16	104	58	120	43	95	56	241	73	223	18	121	299	254	358	207	204	224	190
17	133	87	150	73	123	84	256	88	237	32	151	329	254	13	265	262	283	249
18	162	116	180	103	152	113	271	103	251	46	181	359	269	28	323	320	342	308
19	191	145	210	133	180	141	286	118	265	60	212	30	284	43	21	18	41	7
20	220	174	240	163	209	170	301	133	279	74	242	60	299	58	79	76	100	66
21	249	203	270	193	237	198	316	148	293	88	272	90	314	73	137	134	159	125
22	278	232	300	223	266	227	331	163	307	102	302	120	329	88	195	192	218	184
23	307	261	330	253	294	255	346	178	321	116	332	150	344	103	253	250	277	243

	M_2 3·20 ft.	S_2 1·03 ft.	N_2 0·41 ft.	K_1 0·59 ft.	O_1 0·40 ft.	K_2 0·22 ft.	P_1 0·22 ft.	M_4 0·70 ft.	MS_4 0·47 ft.	A_0	Sum= Height
	ft.	ft.	ft.	ft.	ft.	ft.	ft.	ft.	ft.	ft.	ft.
0	+2·22	+0·24	+0·32	−0·58	−0·36	−0·22	−0·11	+0·70	+0·40	+6·55	9·16
1	+3·06	+0·70	+0·40	−0·53	−0·39	−0·19	−0·15	+0·40	+0·43	+6·55	10·28
2	+3·13	+0·99	+0·39	−0·44	−0·40	−0·11	−0·19	−0·27	+0·05	+6·55	9·70
3	+2·42	+1·00	+0·28	−0·32	−0·38	−0·01	−0·21	−0·69	−0·38	+6·55	8·26
4	+1·10	+0·75	+0·11	−0·18	−0·34	+0·10	−0·22	−0·46	−0·44	+6·55	6·97
5	−0·50	+0·30	−0·09	−0·03	−0·28	+0·19	−0·21	+0·20	−0·07	+6·55	6·06
6	−1·97	−0·24	−0·27	+0·12	−0·21	+0·22	−0·19	+0·68	+0·36	+6·55	5·05
7	−2·94	−0·70	−0·38	+0·27	−0·12	+0·19	−0·16	+0·51	+0·44	+6·55	3·66
8	−3·18	−0·99	−0·40	+0·40	−0·02	+0·11	−0·11	−0·13	+0·10	+6·55	2·33
9	−2·62	−1·00	−0·33	+0·50	+0·07	0·00	−0·06	−0·65	−0·34	+6·55	2·12
10	−1·40	−0·75	−0·17	+0·56	+0·16	−0·11	0·00	−0·56	−0·45	+6·55	3·83
11	+0·17	−0·30	+0·03	+0·59	+0·25	−0·19	+0·05	+0·06	−0·13	+6·55	7·08
12	+1·70	+0·24	+0·22	+0·58	+0·31	−0·22	+0·11	+0·62	+0·33	+6·55	10·44
13	+2·80	+0·70	+0·36	+0·52	+0·37	−0·19	+0·15	+0·60	+0·46	+6·55	12·32
14	+3·20	+0·99	+0·41	+0·43	+0·39	−0·11	+0·19	+0·01	+0·15	+6·55	12·21
15	+2·80	+1·00	+0·35	+0·31	+0·40	0·00	+0·21	−0·59	−0·31	+6·55	10·72
16	+1·70	+0·75	+0·22	+0·17	+0·38	+0·11	+0·22	−0·64	−0·46	+6·55	9·00
17	+0·17	+0·30	+0·04	+0·02	+0·34	+0·19	+0·21	−0·10	−0·17	+6·55	7·55
18	−1·40	−0·24	−0·16	−0·12	+0·28	+0·22	+0·19	+0·53	+0·29	+6·55	6·14
19	−2·62	−0·70	−0·32	−0·28	+0·20	+0·19	+0·16	+0·67	+0·46	+6·55	4·31
20	−3·18	−0·99	−0·40	−0·40	+0·10	+0·11	+0·12	+0·17	+0·19	+6·55	2·27
21	−2·94	−1·00	−0·39	−0·50	+0·01	0·00	+0·06	−0·49	−0·27	+6·55	1·03
22	−1·97	−0·75	−0·28	−0·56	−0·08	−0·11	+0·01	−0·68	−0·47	+6·55	1·66
23	−0·50	−0·30	−0·11	−0·59	−0·17	−0·19	−0·05	−0·24	−0·21	+6·55	4·19

In Example XII, the component times of high water and corrected H of the components have been taken from Example XI, and are given in the first section.

In the second section of the example the solar time is given in the left column, this solar time converted into component time is given in the left column under each component and, in the right column under each component, the difference between the component time and component time of high water, that is to say, I.

In the third section of the example the solar time is again given in the left column, with values of $H \cos I$ under the components, and the sum of all component heights at each hour, which gives the hourly heights required, in the right column.

The hours of solar time are local mean time, and local mean time is always used in calculations by the harmonic method, but may, if required, be converted into standard time after the completion of the calculations.

§ 106. Though the calculations required for finding the height of the tide at each hour of the day by this method appear considerable, it must be remembered that the heights found are, in comparison with those found by other approximate methods, accurate; that the method is applicable to every port in the world; that to find corresponding heights by non-harmonic methods would require the calculation of two Examples II or IV and twenty-four Examples V; that it is not often that heights are required for a whole day; that, in most cases, the components M_4 and MS_4 are not required, and that, though the examples have been worked to hundredths of a foot, sufficient accuracy will be attained in practice if the nearest tenth is used.

§ 107. The times and heights of high and low water may, if required, be obtained approximately from the hourly heights by inspection and are, in Example XII, H.W. 1 h. 15 m., 10·3 ft.; L.W. 8 h. 45 m., 2·1 ft.; H.W. 13 h. 15 m., 12·4 ft.; L.W. 21 h. 15 m., 1·0 ft. If required more exactly the hourly heights must be plotted as in Figure 9, in which figure each component wave is also shown separately.

§ 108. Much information regarding the tides at a port where the harmonic constants are known may be obtained directly from the constants, for:

(*a*) The greatest semi-diurnal tides (springs) will occur $\dfrac{k \text{ of } (S_2 - M_2)}{\text{Speed of } (S_2 - M_2)}$ hours after new and full moon, the least semi-diurnal tides (neaps) at the same interval after quadrature (see § 30).

(*b*) The greatest diurnal tides will occur $\dfrac{k \text{ of } (K_1 - O_1)}{\text{Speed of } (K_1 - O_1)}$ hours after the moon reaches its greatest declination, and the diurnal tide will vanish at the same interval after the moon is on the equator.

(*c*) The possibility and frequency of single day tides depend on the ratio H of $(M_2 - S_2) : H$ of $(K_1 + O_1)$; if H of $(M_2 - S_2)$ is greater than

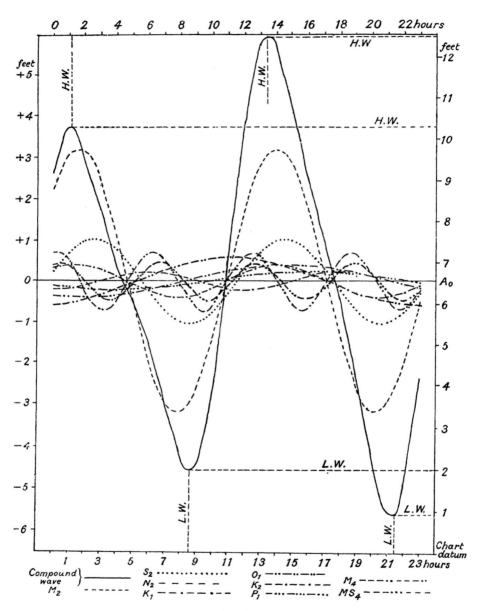

Fig. 9.

H of $(K_1 + O_1)$ single day tides cannot occur; if H of $(K_1 + O_1)$ is just greater than H of $(M_2 - S_2)$ single day tides will occur when the moon is in quadrature at the time of its greatest declination; if H of $(K_1 + O_1)$ is greater than H of $(M_2 + S_2)$ single day tides will always occur when the moon's declination is great and if H of $(K_1 + O_1)$ is very much greater than H of $(M_2 + S_2)$ the tides may become entirely diurnal. The intervals at (a) and (b) above will affect these rules in that the time of the occurrence of single day tides depends on the intervals.

(d) The visible effect of the anomalistic inequality is given by H of N_2, and if, where the tides are of semi-diurnal type, H of N_2 is greater than H of S_2, the anomalistic inequality will exceed the phase inequality and the greatest tides will occur at perigee, not at springs, the least at apogee, not at neaps.

(e) If H of M_4 and MS_4 are appreciable the tides are of shallow water type and the duration of rise and fall unequal.

Etc., etc.

From the above rules, and from the constants given in Table VII:

At Puerto San Antonio the tides are of deep-water semi-diurnal type, but as H of N_2 is equal to H of S_2 the anomalistic inequality is just as marked as the phase inequality; diurnal inequality is small.

At Shanghai the tides are of shallow water semi-diurnal type, but diurnal inequality is considerable.

At Pakhoi the tides are of diurnal type, but as H of M_2 and S_2 are appreciable semi-diurnal tides will probably occur when the moon is near the equator.

At Do Son the tides are almost entirely diurnal; single day tides will probably always occur and will reach their greatest magnitude

$$\frac{91° - 35°}{15°0411 - 13°9430} \text{ hours} = 51 \text{ hours nearly}$$

after the moon reaches its greatest declination and will vanish at the same interval after the moon is on the equator, when there may be semi-diurnal tides of very small magnitude.

Etc., etc.

Note. For exact machine predictions the movement of the imaginary satellites is assumed to be counter-clockwise and k of the components is a minus quantity; this system complicates computation and it has therefore been assumed, in the explanation and description above, that movement is clockwise and k plus.

This change in assumption has necessitated a corresponding change in Tables I and II, which differ from those used for machine predictions, but, provided the same components and constants are used, heights computed by means of the tables will be in exact agreement with those obtained by using the machine.

CHAPTER VIII

Oceanic Currents and Tidal Streams

§ 109. *Oceanic currents* may be either permanent or seasonal; a permanent current runs always in the same, or approximately the same, direction, a seasonal current varies in direction at different seasons or may, at certain seasons, cease running. Currents have no connection with tides and are principally due to prevailing winds, minor causes being differences in rainfall, temperature and salinity in different parts of the ocean; permanent surface currents are due to permanent winds, such as the trade winds, seasonal surface currents to seasonal winds, such as the monsoon; as equilibrium must be maintained surface currents are necessarily accompanied by counter currents, which may be either at or near the surface or at great depths, and a regular circulation of the waters of the ocean is thus established. The principal oceanic surface currents are shown on the Current Charts, prepared and published by the Hydrographic Department of the Admiralty.

Currents must not be confused with, or referred to, as Tidal streams.

The directions of currents are shown on the Admiralty charts either by means of arrows feathered on both sides ($\rangle\!\rangle\!\rangle\!\longrightarrow$) or by unfeathered bent arrows ($\sim\!\sim\!\sim\!\rightarrow$).

Tidal streams are a direct effect of tides, are periodic and may, as the tides, be semi-diurnal or diurnal, but, whereas the tides are periodic vertical movements of the waters on the earth's surface, tidal streams are periodic horizontal movements, and must not, in order that confusion may be avoided, be referred to as tides.

§ 110. Tidal streams are primarily due to differences in relative water level at places not far apart. On the ideal tidal earth described in Chapter I such differences are small, friction would not be overcome and there would be no appreciable streams; this is also the case in the oceans as they exist, but where, owing to decreased depth, the range of the tide wave is increased, the surface slope becomes sufficient for friction to be overcome; local tidal streams are accordingly found on detached shoals in deep water.

§ 111. In the neighbourhood of land, owing to the increased range and decreased speed of the tide wave, surface slope is increased and streams occur which may reach considerable velocity, but, as they usually depend on relative water levels over a large area, neither their direction nor velocity nor their times of slack water can be predicted from theoretical considerations, and it is evident that slack water will not depend at all on the times of local high and low water but on the times when relative levels are such that there is no tendency for the water to run in any direction.

If a perfectly straight channel exist having parallel and vertical sides, and if the depth be the same as that of the tidal sea into which it runs, the range of the tide will gradually decrease as it runs up the channel, and, whether the channel terminates in the land or runs into a land-locked sea not affected by the tides except through the channel:

(a) At high water, sea level will be higher outside than in the channel, and the stream will run in.

(b) At half tide there will be no difference in level, and therefore no stream.

(c) At low water sea level will be lower outside than in the channel, and the stream will run out.

Consequently when sea level is above the mean the stream will run in, when sea level is below the mean the stream will run out, slack water occurring at half tide. These circumstances affect harbours with narrow entrances, channels connecting enclosed bodies of water with the sea, inlets and rivers, though in the latter case the time of slack water is materially changed by the down-running river current.

If the channel connects two bodies of tidal water, the times of slack water and direction and rate of the stream at other times will depend on relative water levels at the ends, slack water occurring when there is no difference of level.

§ 112. In the neighbourhood of land a further cause of tidal streams arises. The motion of the particles of water in a wave in deep water is entirely vertical; when the wave enters shallow water its speed is decreased and height increased and its front slope becomes more, its real slope less, steep; finally, if the depth continues to decrease, the front slope becomes vertical, the wave breaks and the vertical motion of the particles becomes entirely horizontal. Between the vertical motion in deep water and the horizontal motion after the breaking of the wave there is an intermediate stage when the particles have both vertical and horizontal motion. The tide wave seldom breaks, but its shape is changed in shallow water, the particles thus obtain horizontal motion and a tidal stream results. Streams due to this cause run at their maximum velocity at half tide and are slack at high and low water.

In certain rivers, the Severn, Seine, Hugli and Tsien tang kiang for instance, the change in shape of the tide wave, due to shallow water, the configuration of the land, and the down-running river current, is sufficient, under certain astronomical and meteorological conditions, to cause the tide wave to break: the breaking tide wave running up a river is known as a *bore*.

§ 113. The streams as they exist are, almost universally, due to differences in water level (§ 111) combined with the partial breaking of the tide wave (§ 112) and slack water may therefore occur at any time after high water but before half tide and after low water but before half tide. On an open coast it is usually shortly after high and low water, in channels corresponding to that described at § 111 shortly before half tide, but it is never safe to assume a time, which can, like high and low water, only be predicted from observations.

§ 114. Just as the velocity of the stream is increased by shallow water in the oceans, so is its velocity in the neighbourhood of land increased by shoals, uneven bottom, salient points and other obstructions either natural or artificial. Not only does an obstruction increase the rate, but it may also change the direction of that portion of the stream which meets it, and this change in direction may be either horizontal, due to obstructions reaching to the surface, or upward, due to shoals or uneven bottom. Where that portion which has been diverted meets the undiverted main stream, or, in fact, wherever two streams running in different directions meet, *eddies* are formed in which the stream may have a circular motion; when the velocity is great, particularly when the diversion in direction has been upward and when the wind is blowing strongly against the stream, a *race* may occur in which the heavy breaking seas and overfalls thus created are a danger to small vessels.

§ 115. In well defined channels the stream usually follows the direction of the channel; off shore it may be rotary, setting towards a different point of the compass at each hour of the tide. In either case, as the stream depends on the tide, the direction, in the same place, will always be the same at any given interval before or after high water and the velocity will vary with changes in tidal range due to phase and other inequalities.

Owing to friction the tide rises and falls more slowly at the sides of a river or channel than in the centre and more slowly over shoals than in open water; the tide is therefore usually lower when rising and higher when falling at the sides of rivers and channels than in the centre, and over shoals than in open water, and a set towards the sides of rivers and channels and towards shoals may be expected when the tide is rising and away from the sides of rivers and channels and away from shoals when the tide is falling.

§ 116. The stream which runs whilst the tide is rising is sometimes referred to as the *flood* stream, that which runs whilst the tide is falling as the *ebb*. When the stream turns at, or near, the time of high water these terms are unobjectionable, but if the turn of the stream is at about half tide, confusion results from their use; these terms should therefore only be used when the turn of the stream occurs within one hour of the time of high and low water, the stream in other cases being described as *in-going* or *out-going* in rivers, estuaries, etc., or by the *direction towards which it runs*, north-going, east-going, etc., elsewhere.

§ 117. The streams may be predicted, in a similar manner to the tides, either exactly or approximately and harmonically or non-harmonically, from observed directions and rates and times of slack water. The obtaining of a long series of tidal observations is a comparatively simple matter, for observations can be obtained from the shore, a similar series of tidal stream observations, however, requires the continuous employment of a vessel and crew and causes considerable obstruction to traffic, and the observations obtained are therefore not usually of sufficient duration for the calculation of exact predictions. No exact tidal stream predictions are published by the Admiralty, and no information regarding tidal streams is given in the Tide Tables, but the streams are described in the Sailing Directions in which, and on the Charts, information is given by means of which approximate predictions may be calculated; the streams of the British islands, the North sea and North coast of France are also described in a special publication and the streams of the waters surrounding the British islands are shown graphically by means of Tidal Stream Charts.

§ 118. Approximate tidal stream predictions may be calculated either by means of constants or differences, the methods employed being similar to those described in Chapters III and IV; by means of constants the streams are referred directly to the time of the moon's transit, by means of differences they are referred to the time of high water at a port in the neighbourhood; when streams are referred to the time of high water at a Standard port this time should be obtained from the Admiralty Tide Tables, Part I; if the port of reference is not a Standard port the time of high water must be calculated.

The tidal stream information given in Admiralty publications is, owing to the great difficulty and cost of obtaining observations, necessarily incomplete, the calculation of the direction and rate of the stream at any time therefore necessitates several assumptions, and results, when obtained, must consequently be regarded as approximations and used with caution.

§ 119. Several different methods of giving information as to the direction of the tidal stream are in use:

(a) The time when the stream commences to run in a certain direction may be referred to the time of transit of the new or full moon, thus:

"N. going stream commences at 6 h. F. & c.,"

meaning that, on full and change days, the N. going stream commences to run 6 hours after the moon's transit; the interval (6 hours) here corresponds to the H.W.F. & c. and the method of finding the time when the stream commences to run is therefore exactly the same as that of finding the time of high water from the H.W.F. & c. described in Chapter III.

(b) The time when the stream commences to run in a certain direction may be referred to the transit of the moon without qualifications as to phase, thus:

"N. going stream commences at 6 h.,"

meaning that, on the average, the N. going stream commences to run 6 hours after the moon's transit; the interval (6 hours) here corresponds to the mean high water lunitidal interval and the phase inequality to be applied is found by diminishing the time of the moon's transit by 24 minutes for each half day of age of the tide and taking, from Table II (*a*), Admiralty Tide Tables, Part II, the quantity corresponding to the corrected time of transit and age of tide 0 days, the rule thus being: "To the time of the moon's transit, corrected for longitude, apply the interval corrected for phase inequality for age of tide 0 days and time of moon's transit on a day at an interval equal to the age of the tide before the day on which the time of commencement of stream is required."

Note. Methods (*a*) and (*b*) are seldom used; as the observations obtained when these methods are used are never sufficient for the calculation of the phase inequality, it is doubtful whether it is worth considering phase inequality when finding directions at a certain time, and the intervals given may therefore be applied directly to the time of moon's transit on the day.

(*c*) The time when the stream commences to run in a certain direction may be referred to the time of high water, either locally or at another port, thus:

"N. going stream begins 2 h. before H.W. at ——."

The interval (2 hours) here corresponds to a tidal difference on the port of reference, which may be either a Standard port, a Secondary port or a port for which tidal constants only are given. If the port of reference is a Standard port the time of high water must be obtained from the Admiralty Tide Tables, Part I; if the port of reference is a Secondary port the time of high water must be obtained by means of the tidal differences, as described in Chapter IV; if the port of reference is neither a Standard nor a Secondary port the time of high water must be obtained by means of the tidal constants, as described in Chapter III.

(*d*) The direction of the stream may be given at hourly intervals before and after high water, either locally or at another port, and the time of high water at the port of reference must then be found as described at (*c*) above.

§ **120.** Information regarding direction is only complete when given as at § 119 (*d*), and it must, in other cases, be assumed that, unless otherwise stated:

(*a*) The stream continues to run for $6\frac{1}{4}$ hours in the same direction, and

(*b*) The stream then turns through 180° and runs for $6\frac{1}{4}$ hours in the opposite direction.

§ **121.** There are also several different methods in use of giving information as to the velocity of the stream:

(*a*) A single velocity may be given, thus:

"N. going stream commences 2 h. before H.W. at ——, rate 3 kn."

A velocity so given is an average maximum, that is to say, it is the average velocity, half the duration of the stream from its commencement on a day half way between spring and neap tides.

(*b*) A single velocity may be given, thus:

"N. going stream commences 2 h. before H.W. at ——, rate 4 kn. Sp."

A velocity so given is an average maximum spring velocity, that is to say, it is the average velocity, half the duration of the stream from its commencement on the day of spring tides.

(c) A single velocity may be given, thus:

"N. going stream commences 2 h. before H.W. at —, rate 2 kn. Np."

A velocity so given is an average maximum neap velocity, that is to say, it is the average velocity, half the duration of the stream from its commencement on the day of neap tides.

(d) Velocities may be given, thus:

"N. going stream commences 2 h. before H.W. at —, rate 4 kn. Sp., 2 kn. Np."

Velocities so given are average maximum spring and neap velocities as explained at (b) and (c) above.

(e) An average velocity, average spring velocity, average neap velocity or average spring and neap velocities may be given at hourly intervals before and after high water, either locally or at another port.

§ 122. Information regarding velocities is never complete, and velocity at the required time must always be calculated, it being assumed, unless otherwise stated, that:

$$\text{Spring velocity}\quad = 2 \times \text{neap velocity}\quad = \tfrac{4}{3} \times \text{average velocity,} \qquad (122)$$

$$\text{Neap velocity}\quad = \tfrac{1}{2} \times \text{spring velocity} = \tfrac{2}{3} \times \text{average velocity,} \qquad (123)$$

$$\text{Average velocity} = \tfrac{1}{2}\,(\text{spring velocity} + \text{neap velocity})$$

$$= \tfrac{3}{2} \times \text{neap velocity}\quad = \tfrac{3}{4} \times \text{spring velocity,} \qquad (124)$$

$$\text{Velocity on any day} = \text{Spring velocity}$$

$$- \frac{(\text{Spring velocity} - \text{Neap velocity}) \times \text{Days from Spring tide}}{7}, \qquad (125)$$

the days from spring tide being found by expression (53).

Also that, where velocities at hourly intervals before and after high water are not given,

$$\text{Velocity} = V \sin \frac{h \times 180°}{d}, \qquad (126)$$

where V is the velocity, half the duration of the stream from its commencement, h the interval from the commencement of the stream and d the duration of the stream.

Where one velocity only is given it must be assumed that the average maximum velocity for the second period of $6\tfrac{1}{4}$ hours is the same as that given for the first period.

Velocities are occasionally given as "— to — kn." and it must not be assumed that these are neap and spring velocities, the probability being that they are the minimum and maximum velocities observed; in such cases the average of the velocities given should be used as an average maximum velocity.

§ 123. Tidal stream information is sometimes given on charts by means of arrows, as follows:

A plain arrow (———→) indicates directions after high water.

An arrow feathered on one side (⤚⤚⤚→) indicates directions before high water.

Intervals from high water are indicated by means of dots on the arrows, thus —•—•—→ indicates the direction of the stream 3 hours after high water. Velocities in knots and the port of reference are given against the arrows.

Both the direction and velocity and the duration of the stream may be materially influenced by wind; not only does the wind blowing at the time cause changes, but the change in the direction of the stream is not always in the direction of the wind and its effects may persist long after the wind has ceased; for instance, if a long continued south-westerly gale has raised the level of the water in the English Channel above its normal mean level it is reasonable to suppose that, after the force of the south-westerly wind has abated, the west going stream will be stronger than usual until levels have been adjusted. The effects of wind are complex and no general rules can be given.

In Admiralty publications later than the year 1912, where complete tables of tidal streams are given, the directions are true in degrees from 0° to 360°; in the older publications directions are magnetic in degrees from the cardinal points of the compass or in points; in either case directions are approximate only. When a stream is referred to as north going, south-east going, etc., a general direction only is intended. As directions and times of slack water are approximate, times should be calculated to the nearest $\frac{1}{4}$ hour only.

§ 124. In order to facilitate the calculation of velocities, Tables VIII and IX, given in the Appendix, have been computed.

Table VIII gives the value of

$$\frac{(\text{Spring velocity} - \text{Neap velocity}) \times \text{Days from Spring tide}}{7},$$

expression (125); when the spring and neap velocities are not given they may be found, provided either the spring, neap or average velocity is known, from expressions (122) to (124); Table VIII is entered with the difference between the velocities at springs and neaps and the days from spring tide, the corresponding quantity taken out subtracted from the spring velocity giving the velocity on the day in question.

Table IX gives the value of $\sin \dfrac{h \times 180°}{d}$, expression (126), and is entered with the interval from commencement of stream and duration of stream, the corresponding quantity taken out being a factor by which the average maximum velocity of the stream on the day in question, that is to say, the velocity of the stream half its duration after its commencement, must be multiplied in order to obtain the velocity at the time.

Results from the tables are approximate and should be calculated to one place of decimals only.

§ 125. *Example XIII.* If, in a certain position, the only tidal stream information given is "N.E. going stream commences 4 hours after H.W. at Dover, rate 3 kn. Sp." what assumptions may be made?

(*a*) § 120 (*a*), that the stream continues to run to N.E. ward till 10¼ hours after H.W. at Dover = 2¼ hours before the following H.W.

(*b*) § 120 (*b*), that the stream then turns and runs to S.W. ward from 2¼ hours before till 4 hours after H.W. at Dover.

(*c*) § 122, that the velocity given applies both to the N.E. going and the S.W. going streams.

(*d*) § 121 (*b*), that the velocity given is the average velocity half the duration of the stream from its commencement on spring tide day.

(*e*) Expression (123), that the corresponding neap velocity is 1½ kn.

(*f*) That expression (125) will give the corresponding velocity on other days.

(*g*) That expression (126) will give the velocity at times other than half the duration of the stream from its commencement.

Example XIV. If, in a certain position off the east coast of Scotland north of lat. 52° N.:

S.W. going stream begins 3½ hours before H.W. at Dover, rate 2½ kn.

N. ,,   ,,   3½ hours after H.W. ,, ,, 3 ,,

what will be the direction and rate of the stream each hour from 6 h. to 18 h. on 22nd January, 1921?

Admiralty Tide Tables, Part I, H.W. Dover 22nd January, 1921, to nearest ¼ hour, 9 h. 30 m., 21 h. 45 m.

Therefore: S.W. going stream commences at 6 h.
     N. ,, ,, ,, 13 h.
     S.W. ,, ,, ,, 18¼ h.

Admiralty Tide Tables, Part I, Table IV (*b*), Full moon 23rd January, 1921.

Admiralty Tide Tables, Part II, Table I, Age of tide 2 days.

Expression (53), Days from spring tide = 1 day + 2 days = 3 days.

Expression (122),
 Spring velocity of S.W. going stream = ⁴⁄₃ × 2·5 kn. = 3·3 kn.
 Spring velocity of N. going stream = ⁴⁄₃ × 3 kn. = 4 kn.

Expression (123),
 Neap velocity of S.W. going stream = ⅔ × 2·5 kn. = 1·7 kn.
 Neap velocity of N. going stream = ⅔ × 3 kn. = 2 kn.

Therefore, S.W. going, spring velocity − neap velocity
$$= 3\cdot3 \text{ kn.} - 1\cdot7 \text{ kn.} = 1\cdot6 \text{ kn.}$$
N. going, spring velocity − neap velocity
$$= 4 \text{ kn.} - 2 \text{ kn.} = 2 \text{ kn.}$$
Appendix, Table VIII, 3 days from spring tides, difference 1·6 kn.,
$$\text{Velocity} = 3\cdot3 \text{ kn.} - 0\cdot7 \text{ kn.} = 2\cdot6 \text{ kn.}$$
3 days from spring tides, difference 2 kn.,
$$\text{Velocity} = 4\cdot0 \text{ kn.} - 0\cdot9 \text{ kn.} = 3\cdot1 \text{ kn.}$$
Appendix, Table IX, S.W. going stream, duration 7 h., velocity 2·6 kn.:

At 6 h.,	direction of stream S.W.,	velocity	2·6 kn. × 0·0 = 0·0 kn.
,, 7 h.,	,, ,, S.W.,	,,	2·6 kn. × 0·4 = 1·0 kn.
,, 8 h.,	,, ,, S.W.,	,,	2·6 kn. × 0·8 = 2·1 kn.
,, 9 h.,	,, ,, S.W.,	,,	2·6 kn. × 1·0 = 2·6 kn.
,, 10 h.,	,, ,, S.W.,	,,	2·6 kn. × 1·0 = 2·6 kn.
,, 11 h.,	,, ,, S.W.,	,,	2·6 kn. × 0·8 = 2·1 kn.
,, 12 h.,	,, ,, S.W.,	,,	2·6 kn. × 0·4 = 1·0 kn.

N. going stream, duration 5¼ h., velocity 3·1 kn.

At 13 h.,	direction of stream N.,	velocity	3·1 kn. × 0·0 = 0·0 kn.
,, 14 h.,	,, ,, N.,	,,	3·1 kn. × 0·5 = 1·6 kn.
,, 15 h.,	,, ,, N.,	,,	3·1 kn. × 0·9 = 2·8 kn.
,, 16 h.,	,, ,, N.,	,,	3·1 kn. × 1·0 = 3·1 kn.
,, 17 h.,	,, ,, N.,	,,	3·1 kn. × 0·7 = 2·2 kn.
,, 18 h.,	,, ,, N.,	,,	3·1 kn. × 0·2 = 0·6 kn.

CHAPTER IX

Tides and Tidal Streams in Hydrographical Surveying

§ 126. The soundings obtained in the course of a Hydrographical Survey must be reduced to chart datum (see § 34), the level of datum must therefore be determined if not already known.

The reduction of soundings consists in the subtraction from each sounding of the height of the tide above chart datum at the time when it was obtained; the height of the tide must therefore be continuously observed whilst sounding is in progress, and for this purpose a tide pole is required.

§ 127. A graduated plank secured to a pier makes a satisfactory tide pole; if no pier is available a spar must be erected in the sea below the lowest low water level and a graduated plank secured to it, or the spar itself may be graduated. The tide pole should be erected in a position sheltered from the sea but fully open to tidal influence; it should not be in a river, unless the river itself is to be sounded; it should, if possible, be so close to the shore as to be readable without the use of a boat; it should be vertical, or the heights of tide as read will not be accurate; it should be in a position which never dries, and should be of sufficient length to allow of the highest tides being recorded.

The graduation should be such as to be easily read; the best method of graduating is to paint the plank, or spar, in alternate black and white bands, each one foot wide, with numerals, black on white and white on black, 6 inches high, so placed that the bottom of each numeral indicates a complete foot above zero of the pole; the bottom of the figure 10, for instance, will indicate 10·0 ft. above zero, the top 10·5 ft. above zero. Intermediate marks should be made at the quarter feet, and it will then be possible to read the height of sea level above zero of the pole at any moment to within about 0·1 ft. from a considerable distance.

§ 128. After erection, the reading on the tide pole corresponding to chart datum must be found. The Admiralty charts cover the whole world, but information regarding the datum used is, as a rule, only given on those of large scale; where the level of datum is given on the chart, this level must be used, both when a new survey is made and when additional soundings are being obtained, and should be connected to the tide pole as follows:

(*a*) If datum is referred to the top of a pier, or to a mark from which a vertical measurement to water level can be obtained.—Obtain such measurement and a simultaneous reading of water level on the tide pole; add together the measurement and the reading and from their sum subtract the amount which datum is below the mark; the remainder will be the tide pole reading corresponding to chart datum.

(b) If datum is referred to the top of a rock which covers and uncovers.— Obtain a reading of water level on the tide pole when the rock is awash; subtract from this reading the amount which datum is below the top of the rock; the remainder will be the tide pole reading corresponding to chart datum.

(c) If datum is referred to a mark which does not cover and uncover, and from which a vertical measurement to water level cannot be obtained, the difference in level between the mark referred to and a point from which a vertical measurement to water level can be obtained is required. If levelling instruments are available this is a simple matter; if none are available some ingenuity will be required, but, bearing in mind that a line from the observer's eye to the sea horizon, corrected for dip, is horizontal, such difference can usually be obtained by means of a sextant, when proceed as at (a).

§ 129. If the published chart gives no information regarding datum, but if rocks in the vicinity are stated to dry a certain number of feet, an approximate tide pole reading corresponding to chart datum may be obtained as at § 128 (b), but, as the heights of drying rocks are given to the nearest foot only, the mean reading, as obtained from at least three rocks, if available, should be accepted as chart datum.

§ 130. If the published chart gives no information regarding datum and if there are no drying rocks in the vicinity, but if the height of mean level above datum is given in the Admiralty Tide Tables, Part II, the tide pole reading corresponding to mean level should be calculated by summing the heights, observed on the tide pole, of two consecutive high waters and two consecutive low waters and dividing by four; the height of mean level above datum, given in the Tide Tables, subtracted from the calculated tide pole reading of mean level will give approximately the tide pole reading corresponding to chart datum.

Mean level cannot be calculated from less than two consecutive high waters and two consecutive low waters; if time is available additional high and low waters should be observed, but in calculating mean level to ascertain datum the tides must be consecutive, the same number of high and low waters must be used and the number of each must be an even number. The reading on the tide pole corresponding to mean level is always found by summing all the heights and dividing the sum by the total number of high and low waters.

If single day tides occur whilst the observations are in progress this method fails and mean level must be ascertained by summing all the heights observed at hourly intervals for one or more complete lunar days and dividing the sum by the number of heights observed, or mean sea level may be ascertained by the "Challenger" method, described in Wharton and Field's *Hydrographical Surveying*.

§ 131. Where no information regarding datum is given on the published chart, there are no drying rocks in the vicinity and the Tide Tables do not give the height of mean level above datum, the tide pole reading corresponding to chart datum can only be ascertained from observation; this is a matter of considerable difficulty for (see § 34) "Chart datum should be a plane so low that the tide will not often fall below it," and the approximate lowest level to which the tide may fall can only be ascertained from a very long series, a year at least, of observations.

As a rough rule, where the tides are semi-diurnal, chart datum should be about 20 per cent. more than the half mean spring range below mean level and an approximate tide pole reading of this datum may be obtained as follows:

Observe the heights of high and low water day and night for three days; as in § 130, the same number of high and low waters must be observed and each must be an even number; from these observations calculate:

(*a*) Tide pole reading of mean high water (mean of all observed high water heights).

(*b*) Tide pole reading of mean low water (mean of all observed low water heights).

(*c*) Mean range = (M.H.W. − M.L.W.).

(*d*) Mean level = $\frac{1}{2}$ (M.H.W. + M.L.W.).

From the Admiralty Tide Tables, Part I, select a Standard port at which the tides are semi-diurnal and at which the age of the tide is the same as at the place to be surveyed, pick out the heights of high and low water corresponding to the observed heights from which mean level and mean range at the place have been calculated, and calculate mean range at the Standard port.

From the Admiralty Tide Tables, Part I, Table VII or Table VIII, obtain the half mean spring range (Mean Tide Level − Mean Low Water Spring Level) at the Standard port, then, where H = mean level and R = mean range at the place, r = corresponding mean range and r' half mean spring range at the Standard port:

$$\text{Tide pole reading corresponding to chart datum} = H - \frac{12}{10}\left(R \times \frac{r'}{r}\right). \quad (127)$$

Care must be taken that only corresponding tides are used, and, as mean level is not constant and the ratio $m : s$ differs at different places, the resulting level will be approximate only; the three days' observations should, if possible, be obtained from one day before till one day after springs for the resulting datum will then, in all probability, prove satisfactory; if the observations do not include a spring tide the datum found may differ considerably from that required.

§ 132. Where no information regarding datum is given on the published chart, there are no drying rocks in the vicinity, the Tide Tables do not give the height of mean level above datum and the tide is diurnal, or when time will not admit of three days' observations being obtained, the tide pole reading corresponding to mean level should be calculated either by meaning hourly heights for one or more complete lunar days or by the "Challenger" method; the difference between the tide pole readings corresponding to mean level and to the highest high water mark on the beach subtracted from the tide pole reading corresponding to mean level, will give approximately the tide pole reading corresponding to chart datum.

§ 133. It has long been recognised by Hydrographical Surveyors that differences in the chart datums used by different nations are undesirable and that a universal method of ascertaining a universal datum is desirable. These questions were discussed at the International Hydrographic Conference held in London in 1919, and, though the great difficulty of finding a datum from a short series of observations was recognised, the following resolutions were unanimously agreed to:

(i) Tidal datum should be the same as Chart datum, and should be a plane so low that the tide will not frequently fall below it.

(ii) It is greatly to be desired that a uniform datum plane should be adopted by all nations, and the following rule is suggested for the further consideration of hydrographers for a universal datum plane, which should be called "International low water":

"That the plane of reference below mean sea level shall be determined as follows:

"Take half the range between mean lower low water and mean higher high water and multiply this half range by 1·5."

To establish the true level of International low water at any place would require a continuous record of the tides for at least one year, but an approximate value may be obtained from 14 days' observations as follows:

(*a*) From the tide pole readings of the heights of high and low water, pick out the higher high waters and the lower low waters; this can best be done by plotting the heights on section paper, when the high waters will be found to be alternately high and low, the low waters alternately low and high; should single day tides occur, the only high water is the higher high water and the only low water the lower low water.

(*b*) Calculate mean sea level; if hourly heights of tide have been observed this will be the mean of all the hourly heights; if hourly heights have not been observed mean sea level should be calculated by the "Challenger" method for the commencement, middle and end of the observations and a mean of the three values accepted; if neither hourly heights nor the observations required for the "Challenger" method have been obtained the mean height between mean higher high water and mean lower low water must be used.

(*c*) Calculate mean higher high water (mean of all higher high water heights) and mean lower low water (mean of all lower low water heights).

Then, where H is the tide pole reading of mean sea level, R the difference between the tide pole readings of mean higher high water and mean lower low water,

Tide pole reading corresponding to the International low water datum

$$= H - \tfrac{3}{4}R. \tag{128}$$

§ 134. It is probable that "International low water" would be a more satisfactory chart datum than one obtained by the methods described in § 131 and § 132, and it should therefore be used when neither the chart nor the Tide Tables give the information required for finding a datum which has already been in use, and 14 days' continuous tidal observations can be obtained. It is not necessary to wait for the datum to be settled before sounding is commenced, for an arbitrary tide pole reading may be used as a temporary datum to which to reduce the soundings, but the soundings will require correction when the tide pole reading corresponding to chart datum is finally settled; it should be noted that if the arbitrary datum is too high correction to soundings will be minus, if the arbitrary datum is too low correction to soundings will be plus.

When a new datum has been established, whatever the method employed, it must be connected with a fixed mark on shore or a drying rock easily identifiable at a future time, and with the land survey datum if one exists.

§ 135. If a few soundings, for the correction of the published chart only, are to be obtained, then, after the observations required for the datum have been obtained, it will only be necessary to observe the tides whilst sounding is actually in progress; if, however, a complete survey of the place is to be undertaken, and if it is considered that this will last 14 days or more, every effort should be made to observe the tides day and night continuously in order that constants may be calculated, for which purpose continuous observations obtained for only one half lunation are to be preferred to a much longer series observed by day only.

No attempt need be made to calculate constants, but the observations, which should include the heights of the tide at hourly intervals when movement is rapid and at intervals of ten minutes from about half an hour before till half an hour after high and low water, should be sent to the Hydrographic Department of the Admiralty or other authority publishing charts and tide tables for the locality. The tidal record should be complete in itself and should therefore include the day of the month, month and year, the name of the place and its latitude and longitude, the exact local position of the tide pole, the tide pole reading corresponding to chart datum, connection between chart datum and fixed mark and land survey datum, if one exists, and whether time kept was standard or local.

§ 136. When a survey covers a large area more tidal stations than one will be required; when this is the case the tides should be observed continuously at one central station only, other stations, at which the tides need only be observed when required for the reduction of soundings, being regarded as subsidiary.

§ 137. When tides· are observed at more stations than one it is desirable, unless the range of the tide differs considerably, that the same plane should be used as chart datum at all. Chart datum should be transferred from station to station by means of the land survey datum where one exists; if there is no land survey datum, it cannot be assumed that water level forms a plane surface either at low water or at any other moment, and datum should therefore not be transferred from station to station by means of single tide pole readings obtained at low water or at any other time, but it must be assumed that mean level is a plane surface and datum should therefore be transferred as follows:

Obtain a few days' continuous observations at the central and subsidiary stations and, using corresponding observations only, calculate mean level at both; then, where H is the reading of mean level on the subsidiary pole, d the difference between the readings of mean level and chart datum on the central pole,

Reading on subsidiary pole corresponding to chart datum $= H - d.$ (129)

Datum at each subsidiary tidal station must be ascertained by connection with the central station, or with the land survey datum if one exists.

§ 138. When tides are observed at more stations than one and it is seen that considerable differences exist in tidal range, chart datum should not be a plane surface, but the distance between mean level and chart datum at each tidal station should be proportionate to the range, and chart datum at the subsidiary stations should be obtained as follows:

Obtain a few days' continuous observations at the central and subsidiary stations and, using corresponding tides only, calculate mean high water level, mean low water level, mean range and mean level at each; then, where H = reading of mean level on subsidiary pole, d = difference between readings of mean level and chart datum on central pole, R = mean range at subsidiary station and r = mean range at central station,

Reading on subsidiary pole corresponding to chart datum $= H - d\dfrac{R}{r}.$ (130)

Datum at each subsidiary station must be ascertained by reference to the central station, and should be connected with an easily identifiable mark on shore or rock which covers and uncovers and with the land survey datum at all stations.

§ **139.** The tidal streams should be observed in positions where their direction, velocity and time of slack water are of importance to navigation. The observations should be made, as far as possible, at exact hourly intervals before and after local high water, or before and after high water at a Standard port in the vicinity; where the time of local high water is not known and there is no Standard port in the vicinity, observations should be obtained at exact hourly intervals before and after the moon's transit; they must be continued for half a lunar day at least where the tide is entirely, or almost entirely, semi-diurnal, but where diurnal inequality is great or single day tides occur they must be continued for at least one complete lunar day; when possible 14 complete days' observations should be obtained.

The observations should be obtained from a vessel or boat at anchor by means of a log-ship, of ordinary pattern but of large size, or a light pole, weighted to float in an upright position, with fans projecting just below its water line, and a light line, marked at intervals of 10 feet, and secured to the log-ship or pole by a span; the stray line should be sufficient to take the log-ship clear of currents due to the vessel. A stop watch should be used if available and, as streams are usually comparatively weak, the line should be allowed to run out for some minutes; a good rule is always to allow 100 ft. of line to run out unless this takes more than 3 minutes, if the stream is very weak the line may be stopped at 3 minutes. The direction of the stream should be noted by means of a compass at the spot where the observations are obtained, for if the observations are obtained at the stern of a vessel at anchor, but swung across the stream by wind, the direction as observed from the standard compass may be seriously in error.

As there are 6080 ft. in a nautical mile and 60 minutes in an hour,

$$\text{Rate of stream in knots} = \frac{\text{Line run out} \times 60}{\text{Number of minutes} \times 6080}, \tag{131}$$

or, more approximately,

$$\text{Rate of stream in knots} = \frac{\text{Line run out}}{\text{Number of minutes} \times 100}. \tag{132}$$

Velocities should be observed, not estimated, for estimated velocities are almost invariably too great. The direction of the stream and time of slack water may, however, be obtained from a vessel at anchor in calm weather by noting the direction of the ship's head and time when she swings.

The direction and force of the wind should be noted at each observation.

§ **140.** The surface stream is much influenced, both in direction and velocity, by wind; the sub-surface stream is not affected to the same extent and, even apart from wind effects, may differ from the surface stream. The sub-surface stream should therefore also be observed, particularly when the observations are restricted to one-half lunar day and are therefore not very reliable. Special instruments have been designed for observing; if such are not available, efficient substitutes may be constructed on the principle of exposing an object of large superficial area to the sub-surface stream and supporting this object by a buoy of small superficial area on the surface.

A suitable instrument for observing consists of a square wooden board measuring some 3 or 4 feet each way, weighted at one edge so as to sink in an upright position, suspended from a surface buoy of just sufficient size to support it; the line, which should be heavier than that used for the surface observations but not heavier than necessary, is secured to the board by means of a span in the same manner as the surface log-ship, or weighted pole, and observations are obtained in the same manner as the surface observations. The most useful depth at which to obtain observations is at from 3 to 5 fathoms; the board will not remain vertically under the buoy, unless the direction and velocity of the sub-surface stream is exactly the same as that of the surface stream, and the distance from the buoy to the centre of the board should therefore be somewhat greater than the depth at which observations are desired; a distance of 6 fathoms will almost certainly ensure the observations being within the required depth limits. The drift of the board will be the resultant of the action of the surface stream on the buoy and the sub-surface stream on the board, but, provided the buoy is small, error may be neglected.

§ 141. A record of tidal stream observations should be sent to the Hydrographic Department of the Admiralty or other authority publishing charts and tide tables for the locality; this should be complete in itself and should therefore include, in addition to the observations, the day of the month, month and year, the name of the place and its latitude and longitude, the exact position of the observations either by sextant angles or bearing and distance of points of land, direction and force of the wind at each observation, depth at which observations were obtained if sub-surface, time of local high water or other phenomena to which the observations are referred, whether time kept is standard or local, and whether true or magnetic stream directions are given.

APPENDIX

TABLE I. Giving the astronomical arguments of the components for the meridian of Greenwich at 0 h. 00 m. G.M.T. on 1st January. (Midnight 31st December—1st January.)

Year	M_2 and MS_4	S_2	N_2	K_1	O_1	K_2	P_1	M_4
1920	**235°**		**204°**	**343°**	**254°**	**149°**		**109°**
1921	159		230	345	175	152		318
22	59		218	350	70	160		118
23	319		207	352	324	165		278
24	**219**		**196**	**356**	**219**	**172**		**78**
25	143		221	356	142	173		286
1926	43		210	358	39	177		85
27	302		198	359	298	179		244
28	**201**		**186**	**358**	**198**	**178**		**42**
29	125		211	356	124	174		249
30	23		201	355	28	171		46
1931	282		186	352	288	166		203
32	**180**	Invariably 0°	**173**	**350**	**189**	**162**	Invariably 10°	**0**
33	103		198	347	117	155		206
34	2		185	345	18	150		3
35	261		173	343	280	146		161
1936	**160**		**160**	**342**	**181**	**144**		**319**
37	83		185	340	106	142		167
38	343		174	341	5	143		326
39	242		162	343	262	147		125
40	**142**		**151**	**346**	**158**	**153**		**285**
1941	67		177	349	78	159		134
42	327		166	352	332	165		294
43	227		154	353	227	171		94
44	**126**		**143**	**358**	**124**	**176**		**253**
45	50		169	358	47	177		101
1946	310		157	358	306	178		259
47	209		145	358	207	177		57
48	**108**		**132**	**356**	**108**	**174**		**215**
49	31		157	353	35	168		61
50	289		145	351	297	163		218

Astronomical arguments for Leap years are in heavy type.

TABLE II. Correction for longitude: East longitude subtract, West longitude add.

Component	15°	30°	45°	60°	75°	90°	105°	120°	135°	150°	165°	180°
M_2, O_1 and MS_4	1°	2°	3°	4°	5°	6°	7°	8°	9°	10°	11°	12°
N_2	2	3	5	7	8	9	11	12	14	16	17	19
M_4	2	4	6	8	10	12	14	16	18	20	22	24

TABLE III. Giving the increment to the astronomical argument at 0 h. on 1st January for 0 h. on each day of the year.

(a) Components M_2, hourly speed 28°9842, and MS_4, hourly speed 58°9842.

(b) Component S_2, hourly speed 30°0000.

(b)

Month day	Jan.	Feb.	Mar.	Ap.	May	June	July	Aug.	Sep.	Oct.	Nov.	Dec.
1												
2												
3												
4												
5												
6												
7						Invariably 0° at 0 h.						
8												
9												
10												
11												
12												
13												
14												
15												
16												
17												
18												
19												
20												
21												
22												
23												
24												
25												
26												
27												
28												
29												
30												
31												

(a)

Month day	Jan.	Feb.	Mar.	Ap.	May	June	July	Aug.	Sep.	Oct.	Nov.	Dec.
1	0°	36°	359°	34°	46°	82°	93°	129°	165°	176°	212°	224°
2	24	60	23	59	70	106	117	153	189	201	237	248
3	49	85	47	83	95	130	142	178	213	225	261	272
4	73	109	72	107	119	155	166	202	238	249	285	297
5	98	133	96	132	143	179	191	226	262	274	310	321
6	122	158	120	156	168	204	215	251	287	298	334	346
7	146	182	145	181	192	228	239	275	311	323	359	10
8	171	206	169	205	216	252	264	299	335	347	23	34
9	195	231	194	229	241	277	288	324	0	11	47	59
10	219	255	218	254	265	301	312	348	24	36	72	83
11	244	280	242	278	290	325	337	13	49	60	96	107
12	268	304	267	303	314	350	1	37	73	85	120	132
13	293	328	291	327	338	14	26	61	97	109	145	156
14	317	353	315	351	3	39	50	86	122	133	169	181
15	341	17	340	16	27	63	74	110	146	158	194	205
16	6	42	4	40	52	87	99	135	170	182	218	229
17	30	66	29	64	76	112	123	159	195	206	242	254
18	54	90	53	89	100	136	148	183	219	231	267	278
19	79	115	77	113	125	160	172	208	244	255	291	303
20	103	139	102	138	149	185	196	232	268	280	315	327
21	128	163	126	162	173	209	221	257	293	304	340	351
22	152	188	151	186	198	234	245	281	317	328	4	16
23	176	212	175	211	222	258	269	305	341	353	29	40
24	201	237	199	235	247	282	294	330	6	17	53	64
25	225	261	224	259	271	307	318	354	30	42	77	89
26	250	285	248	284	295	331	343	18	54	66	102	113
27	274	310	272	308	320	356	7	43	79	90	126	138
28	298	334	297	333	344	20	31	67	103	115	151	162
29	323	.	321	357	8	44	56	92	128	139	175	186
30	347	.	346	21	33	69	80	116	152	163	199	211
31	11	.	10	.	57	.	104	140	.	188	.	235
												259

In Leap years add one day to dates after 28 February.

TABLE III (*cont.*). Giving the increment to the astronomical argument at 0 h. on 1st January for 0 h. on each day of the year.

(*c*) Component N_2, hourly speed 28°·4398.

Month day	Jan.	Feb.	Mar.	Ap.	May	June	July	Aug.	Sep.	Oct.	Nov.	Dec.
1	0°	81°	49°	130°	174°	255°	298°	19°	100°	143°	224°	268°
2	37	118	87	168	211	292	335	56	137	181	262	305
3	75	156	124	205	248	330	13	94	175	218	299	342
4	112	193	162	243	286	7	50	131	212	256	336	20
5	150	231	199	280	323	44	88	169	249	293	14	57
6	187	268	237	317	1	82	125	206	287	331	51	95
7	225	306	274	355	38	119	163	244	324	8	89	132
8	262	343	311	32	76	157	200	281	2	45	126	170
9	300	20	349	70	113	194	238	318	39	83	164	207
10	337	58	26	107	151	232	275	356	77	120	201	245
11	14	95	64	145	188	269	312	33	114	158	239	282
12	52	133	101	182	226	307	350	71	152	195	276	319
13	89	170	139	220	263	344	27	108	189	233	313	357
14	127	208	176	257	301	21	65	146	227	270	351	34
15	164	245	214	294	338	59	102	183	264	308	28	72
16	202	283	251	332	15	96	140	221	302	345	66	109
17	239	320	288	9	53	134	177	258	339	22	103	147
18	277	357	326	47	90	171	215	295	16	60	141	184
19	314	35	3	84	128	209	252	332	54	97	178	222
20	351	72	41	122	165	246	289	10	91	135	216	259
21	29	110	78	159	203	284	327	48	129	172	253	296
22	66	147	116	197	240	321	4	85	166	210	290	333
23	104	185	153	234	278	358	42	123	204	247	328	11
24	141	222	191	271	315	36	79	160	241	285	5	49
25	179	260	228	309	352	73	117	198	279	322	43	86
26	216	297	266	346	30	111	154	235	316	359	80	124
27	254	334	303	24	67	148	192	272	353	37	118	161
28	291	12	340	61	105	186	229	310	31	74	155	199
29	329	.	18	99	142	223	267	347	68	112	193	236
30	6	.	55	136	180	261	304	25	105	149	230	273
31	43	.	93	.	217	.	341	62	.	187	.	311

In Leap years add one day to all dates after 28 February.

(*d*) Component K_1, hourly speed 15°·0411.

Month day	Jan.	Feb.	Mar.	Ap.	May	June	July	Aug.	Sep.	Oct.	Nov.	Dec.
1	0°	329°	302°	271°	242°	211°	181°	151°	120°	90°	60°	30°
2	359	328	301	270	241	210	180	150	119	89	59	29
3	358	327	300	269	240	209	179	149	118	88	58	28
4	357	326	299	268	239	208	178	148	117	87	57	27
5	356	325	298	267	238	207	177	147	116	86	56	26
6	355	324	297	266	237	206	176	146	115	85	55	25
7	354	323	296	265	236	205	175	145	114	84	54	24
8	353	322	295	264	235	204	174	144	113	83	53	23
9	352	321	294	263	234	203	173	143	112	82	52	22
10	351	320	293	262	233	202	172	142	111	81	51	21
11	350	319	292	261	232	201	171	141	110	80	50	21
12	349	318	291	260	231	200	170	140	109	79	49	20
13	348	317	290	259	230	199	169	139	108	78	48	19
14	347	316	289	258	229	198	168	138	107	77	47	18
15	346	315	288	257	228	197	167	137	106	76	46	17
16	345	315	287	256	227	196	166	136	105	76	45	16
17	344	314	286	255	226	195	165	135	104	75	44	15
18	343	313	285	254	225	194	164	134	103	74	43	14
19	342	312	284	253	224	193	163	133	102	73	42	13
20	341	311	283	252	223	192	162	132	101	72	41	12
21	340	310	282	251	222	191	162	131	100	71	40	11
22	339	309	281	250	221	190	161	130	99	70	39	10
23	338	308	280	249	220	189	160	129	98	69	38	9
24	337	307	279	248	219	188	159	128	97	68	37	8
25	336	306	278	247	218	187	158	127	96	67	36	7
26	335	305	277	246	217	186	157	126	95	66	35	6
27	334	304	276	245	216	185	156	125	94	65	34	5
28	333	303	275	244	215	184	155	124	93	64	33	4
29	332	.	274	243	214	183	154	123	92	63	32	3
30	331	.	273	242	213	182	153	122	91	62	31	2
31	330	.	272	.	212	.	152	121	.	61	.	0

In Leap years add one day to all dates after 28 February.

TABLE III (cont.). Giving the increment to the astronomical argument at 0 h. on 1st January for 0 h. on each day of the year.

(f) Component K_2, hourly speed 30°·0822.

Month day	Jan.	Feb.	Mar.	Ap.	May	June	July	Aug.	Sep.	Oct.	Nov.	Dec.
1	0°	299°	244°	183°	123°	62°	3°	302°	241°	182°	121°	61°
2	358	297	242	181	122	60	1	300	239	180	119	60
3	356	295	240	179	120	58	359	298	237	178	117	58
4	354	293	238	177	118	56	357	296	235	176	115	56
5	352	291	236	175	116	54	355	294	233	174	113	54
6	350	289	234	173	114	52	353	292	231	172	111	52
7	348	287	232	171	112	51	351	290	229	170	109	50
8	346	285	230	169	110	49	349	288	227	168	107	48
9	344	283	228	167	108	47	347	286	225	166	105	46
10	342	281	226	165	106	45	345	284	223	164	103	44
11	340	279	224	163	104	43	343	282	221	162	101	42
12	338	277	222	161	102	41	342	280	219	160	99	40
13	336	275	220	159	100	39	340	278	217	158	97	38
14	334	273	218	157	98	37	338	276	215	156	95	36
15	332	271	216	155	96	35	336	274	213	154	93	34
16	330	269	214	153	94	33	334	272	211	152	91	32
17	328	267	212	151	92	31	332	271	209	150	89	30
18	327	265	210	149	90	29	330	269	207	148	87	28
19	325	263	208	147	88	27	328	267	205	146	85	26
20	323	261	206	145	86	25	326	265	203	144	83	24
21	321	259	204	143	84	23	324	263	201	142	81	22
22	319	257	202	141	82	21	322	261	200	140	79	20
23	317	256	200	139	80	19	320	259	198	138	77	18
24	315	254	198	137	78	17	318	257	196	136	75	16
25	313	252	196	135	76	15	316	255	194	134	73	14
26	311	250	194	133	74	13	314	253	192	132	71	12
27	309	248	192	131	72	11	312	251	190	131	69	10
28	307	246	191	129	70	9	310	249	188	129	67	8
29	305	.	189	127	68	7	308	247	186	127	65	6
30	303	.	187	125	66	5	306	245	184	125	63	4
31	301	.	185	.	64	.	304	243	.	123	.	2

In Leap years add one day to dates after 28 February.

(e) Component O_1, hourly speed 13°·9431.

Month day	Jan.	Feb.	Mar.	Ap.	May	June	July	Aug.	Sep.	Oct.	Nov.	Dec.
1	0°	66°	57°	123°	164°	230°	271°	338°	44°	85°	151°	192°
2	25	92	82	148	189	256	297	3	69	110	177	218
3	51	117	107	174	215	281	322	28	95	136	202	243
4	76	142	133	199	240	306	347	54	120	161	227	268
5	101	168	158	224	265	332	13	79	145	186	253	294
6	127	193	183	250	291	357	38	104	171	212	278	319
7	152	219	209	275	316	23	64	130	196	237	304	345
8	178	244	234	301	342	48	89	155	222	263	329	10
9	203	269	260	326	7	73	114	181	247	288	354	35
10	228	295	285	351	32	99	140	206	272	313	20	61
11	254	320	310	17	58	124	165	231	298	339	45	86
12	279	345	336	42	83	149	190	257	323	4	70	111
13	304	11	1	67	108	175	216	282	348	29	96	137
14	330	36	26	93	134	200	241	307	14	55	121	162
15	355	62	52	118	159	225	266	333	39	80	146	187
16	21	87	77	143	184	251	292	358	65	105	172	213
17	46	112	102	169	210	276	317	24	90	131	197	238
18	71	138	128	194	235	302	343	49	115	156	223	264
19	97	163	153	220	261	327	8	74	141	182	248	289
20	122	188	179	245	286	352	33	100	166	207	273	314
21	147	214	204	270	311	18	59	125	191	232	299	340
22	173	239	229	296	337	43	84	150	217	258	324	5
23	198	264	255	321	2	68	109	176	242	283	349	30
24	223	290	280	346	27	94	135	201	267	308	15	56
25	249	315	305	12	53	119	160	226	293	334	40	81
26	274	341	331	37	78	144	185	252	318	359	66	106
27	300	6	356	63	103	170	211	277	344	25	91	132
28	325	31	22	88	129	195	236	303	9	50	116	157
29	350	.	47	113	154	221	262	328	34	75	142	183
30	16	.	72	139	180	246	287	353	60	101	167	208
31	41	.	98	.	205	.	312	19	.	126	.	233

In Leap years add one day to dates after 28 February.

TABLE III (cont.). Giving the increment to the astronomical argument at 0 h. on 1st January for 0 h. on each day of the year.

(h) Component M_4, hourly speed 57°·9684.

Month day	Jan.	Feb.	Mar.	Ap.	May	June	July	Aug.	Sep.	Oct.	Nov.	Dec.
1	0°	72°	357°	69°	91°	163°	186°	258°	330°	353°	64°	87°
2	49	120	46	117	140	212	235	307	19	42	113	136
3	98	169	95	166	189	260	283	356	67	90	162	185
4	146	218	143	215	238	309	332	45	116	139	211	234
5	195	267	192	264	286	358	21	93	165	188	259	282
6	244	315	241	312	335	47	70	142	214	237	308	331
7	293	4	290	1	24	96	118	191	262	285	357	20
8	341	53	338	50	73	144	167	240	311	334	46	69
9	30	102	27	99	121	193	216	288	0	23	95	117
10	79	151	76	147	170	242	265	337	49	72	143	166
11	128	199	125	196	219	291	313	26	98	120	192	215
12	176	248	173	245	268	339	2	75	146	169	241	264
13	225	297	222	294	316	28	51	123	195	218	290	312
14	274	346	271	342	5	77	100	172	244	267	338	1
15	323	34	320	31	54	126	148	221	293	315	27	50
16	11	83	8	80	103	174	197	270	341	4	76	99
17	60	132	57	129	152	223	246	318	30	53	125	147
18	109	181	106	177	200	272	295	7	79	102	173	196
19	158	229	155	226	249	321	343	56	128	151	222	245
20	206	278	203	275	298	9	32	105	176	199	271	294
21	255	327	252	324	347	58	81	154	225	248	320	342
22	304	16	301	12	35	107	130	202	274	297	8	31
23	353	64	350	61	84	156	178	251	323	346	57	80
24	42	113	39	110	133	204	227	300	11	34	106	129
25	90	162	87	159	182	253	276	349	60	83	155	177
26	139	211	136	207	230	302	325	37	109	132	203	226
27	188	259	185	256	279	351	13	86	158	181	252	275
28	237	308	234	305	328	40	62	135	206	229	301	324
29	285	·	282	354	17	88	111	184	255	278	350	12
30	334	·	331	43	65	137	160	232	304	327	39	61
31	23	·	20	·	114	·	208	281	·	16	·	110

In Leap years add one day to dates after 28 February.

(g) Component P_1, hourly speed 14°·9589.

Month day	Jan.	Feb.	Mar.	Ap.	May	June	July	Aug.	Sep.	Oct.	Nov.	Dec.
1	0°	31°	58°	89°	119°	149°	179°	209°	240°	270°	300°	330°
2	1	32	59	90	119	150	180	210	241	270	301	331
3	2	33	60	91	120	151	181	211	242	271	302	332
4	3	34	61	92	121	152	182	212	243	272	303	333
5	4	35	62	93	122	153	183	213	244	273	304	334
6	5	36	63	94	123	154	184	214	245	274	305	335
7	6	37	64	95	124	155	185	215	246	275	306	336
8	7	38	65	96	125	156	186	216	247	276	307	337
9	8	39	66	97	126	157	187	217	248	277	308	338
10	9	40	67	98	127	158	188	218	249	278	309	339
11	10	41	68	99	128	159	189	219	250	279	310	340
12	11	42	69	100	129	160	190	220	251	280	311	341
13	12	43	70	101	130	161	191	221	252	281	312	342
14	13	44	71	102	131	162	192	222	253	282	313	343
15	14	45	72	103	132	163	193	223	254	283	314	344
16	15	45	73	104	133	164	193	224	255	284	315	345
17	16	46	74	105	134	165	194	225	256	285	316	346
18	17	47	75	106	135	166	195	226	257	286	317	347
19	18	48	76	107	136	167	196	227	258	287	318	348
20	19	49	77	108	137	168	197	228	259	288	319	349
21	20	50	78	109	138	169	198	229	260	289	320	350
22	21	51	79	110	139	170	199	230	261	290	321	351
23	22	52	80	111	140	171	200	231	262	291	322	352
24	23	53	81	112	141	172	201	232	263	292	323	353
25	24	54	82	113	142	173	202	233	264	293	324	354
26	25	55	83	114	143	174	203	234	265	294	325	355
27	26	56	84	115	144	175	204	235	266	295	326	356
28	27	57	85	116	145	176	205	236	267	296	327	357
29	28	·	86	117	146	177	206	237	268	297	328	358
30	29	·	87	118	147	178	207	238	269	298	329	359
31	30	·	88	·	148	·	208	239	·	299	·	0

In Leap years add one day to dates after 28 February.

TABLE IV.

Giving intervals of mean solar time in component time.

Hour	M_2	S_2	N_2	K_1	O_1	K_2	P_1	M_4	MS_4
0	0°	0°	0°	0°	0°	0°	0°	0°	0°
1	29	30	28	15	14	30	15	58	59
2	58	60	57	30	28	60	30	116	118
3	87	90	85	45	42	90	45	174	177
4	116	120	114	60	56	120	60	232	236
5	145	150	142	75	70	150	75	290	295
6	174	180	171	90	84	180	90	348	354
7	203	210	199	105	98	211	105	46	53
8	232	240	228	120	112	241	120	104	112
9	261	270	256	135	125	271	135	162	171
10	290	300	284	150	139	301	150	220	230
11	319	330	313	165	153	331	165	278	289
12	348	0	341	180	167	1	180	336	348
13	17	30	10	196	181	31	194	34	47
14	46	60	38	211	195	61	209	92	106
15	75	90	67	226	209	91	224	150	165
16	104	120	95	241	223	121	239	207	224
17	133	150	123	256	237	151	254	265	283
18	162	180	152	271	251	181	269	323	342
19	191	210	180	286	265	212	284	21	41
20	220	240	209	301	279	242	299	79	100
21	249	270	237	316	293	272	314	137	159
22	278	300	266	331	307	302	329	195	218
23	307	330	294	346	321	332	344	253	277
24	336	0	323	1	335	2	359	311	336

TABLE V.

Giving the factor for H of the components according to the year.

Year	M_2 and N_2	K_1	O_1	K_2
1920	1·03	0·92	0·87	0·81
1921	1·03	0·89	0·83	0·77
22	1·04	0·88	0·80	0·75
23	1·04	0·89	0·82	0·76
24	1·03	0·91	0·85	0·79
25	1·02	0·94	0·91	0·85
1926	1·01	0·98	0·97	0·94
27	1·00	1·02	1·03	1·03
28	0·99	1·06	1·09	1·13
29	0·98	1·08	1·14	1·22
30	0·97	1·10	1·17	1·28
1931	0·97	1·11	1·18	1·31
32	0·96	1·11	1·18	1·31
33	0·97	1·10	1·16	1·27
34	0·98	1·08	1·13	1·20
35	0·99	1·05	1·08	1·12
1936	1·00	1·02	1·03	1·02
37	1·01	0·98	0·96	0·93
38	1·02	0·94	0·90	0·85
39	1·03	0·91	0·85	0·78
40	1·04	0·89	0·81	0·75
1941	1·04	0·88	0·80	0·75
42	1·03	0·89	0·83	0·77
43	1·02	0·92	0·87	0·81
44	1·01	0·96	0·93	0·88
45	1·00	1·00	0·99	0·97
1946	0·99	1·03	1·05	1·07
47	0·98	1·07	1·11	1·16
48	0·97	1·09	1·15	1·24
49	0·96	1·11	1·17	1·30
50	0·96	1·11	1·18	1·32

TABLE VI.

Giving value of $H \cos I$, where H is the semi-amplitude of the component and I the difference, in component time, between the time of high water of the partial wave and the time at which its height is required.

I \ H	1·00	2·00	3·00	4·00	5·00	6·00	7·00	8·00	9·00	10·00
°	ft.	ft.	ft.	ft.	ft.	ft.	ft.	ft.	ft.	ft.
1	1·00	2·00	3·00	4·00	5·00	6·00	7·00	8·00	9·00	10·00
2	1·00	2·00	3·00	4·00	5·00	5·99	6·99	7·99	8·99	9·99
3	1·00	2·00	3·00	4·00	5·00	5·99	6·99	7·99	8·99	9·99
4	1·00	2·00	2·99	3·99	4·99	5·98	6·98	7·98	8·98	9·98
5	0·99	1·99	2·99	3·98	4·98	5·97	6·97	7·96	8·96	9·96
6	0·99	1·99	2·98	3·98	4·97	5·97	6·96	7·96	8·95	9·95
7	0·99	1·99	2·98	3·97	4·96	5·96	6·95	7·94	8·94	9·93
8	0·99	1·98	2·97	3·96	4·95	5·94	6·93	7·92	8·91	9·90
9	0·99	1·97	2·96	3·95	4·94	5·93	6·92	7·90	8·89	9·88
10	0·98	1·97	2·95	3·94	4·92	5·91	6·89	7·88	8·86	9·85
11	0·98	1·96	2·95	3·93	4·91	5·89	6·87	7·86	8·84	9·82
12	0·98	1·96	2·93	3·91	4·89	5·87	6·85	7·82	8·80	9·78
13	0·97	1·95	2·92	3·90	4·87	5·84	6·82	7·79	8·77	9·74
14	0·97	1·94	2·91	3·88	4·85	5·82	6·79	7·76	8·73	9·70
15	0·97	1·93	2·90	3·86	4·83	5·80	6·76	7·73	8·69	9·66
16	0·96	1·92	2·88	3·84	4·81	5·77	6·73	7·69	8·65	9·61
17	0·96	1·91	2·87	3·82	4·78	5·74	6·70	7·65	8·60	9·56
18	0·95	1·90	2·85	3·80	4·75	5·71	6·66	7·61	8·56	9·51
19	0·95	1·89	2·84	3·78	4·73	5·68	6·62	7·57	8·51	9·46
20	0·94	1·88	2·82	3·76	4·70	5·64	6·58	7·52	8·46	9·40
21	0·93	1·87	2·80	3·74	4·67	5·60	6·54	7·47	8·41	9·34
22	0·93	1·85	2·78	3·71	4·64	5·56	6·49	7·42	8·34	9·27
23	0·92	1·84	2·76	3·68	4·61	5·53	6·45	7·37	8·29	9·21
24	0·91	1·83	2·74	3·66	4·57	5·48	6·40	7·31	8·23	9·14
25	0·91	1·81	2·72	3·63	4·53	5·44	6·34	7·25	8·17	9·06
26	0·90	1·80	2·70	3·60	4·50	5·39	6·29	7·19	8·09	8·99
27	0·89	1·78	2·67	3·56	4·46	5·35	6·24	7·13	8·02	8·91
28	0·88	1·77	2·65	3·53	4·42	5·30	6·18	7·06	7·95	8·83
29	0·88	1·75	2·63	3·50	4·38	5·25	6·13	7·00	7·88	8·75
30	0·87	1·73	2·60	3·46	4·33	5·20	6·06	6·93	7·79	8·67
31	0·86	1·71	2·57	3·42	4·29	5·14	6·00	6·86	7·71	8·57
32	0·85	1·70	2·54	3·39	4·24	5·09	5·94	6·78	7·63	8·48
33	0·84	1·68	2·52	3·36	4·20	5·03	5·87	6·71	7·55	8·39
34	0·83	1·66	2·49	3·32	4·15	4·97	5·80	6·63	7·46	8·29
35	0·82	1·64	2·46	3·28	4·10	4·92	5·73	6·55	7·37	8·19
36	0·81	1·62	2·43	3·24	4·05	4·85	5·66	6·47	7·28	8·09
37	0·80	1·60	2·40	3·20	4·00	4·79	5·59	6·39	7·19	7·99
38	0·79	1·58	2·36	3·15	3·94	4·73	5·52	6·30	7·09	7·88
39	0·78	1·55	2·33	3·11	3·89	4·66	5·44	6·22	6·99	7·77
40	0·77	1·53	2·30	3·05	3·83	4·60	5·34	6·12	6·89	7·66
41	0·76	1·51	2·27	3·02	3·78	4·53	5·29	6·04	6·80	7·55
42	0·74	1·49	2·23	2·97	3·72	4·46	5·20	5·94	6·69	7·43
43	0·73	1·46	2·19	2·92	3·66	4·39	5·12	5·85	6·58	7·31
44	0·72	1·44	2·16	2·88	3·60	4·31	5·03	5·75	6·47	7·19
45	0·71	1·41	2·12	2·83	3·54	4·24	4·95	5·66	6·36	7·07

Where I is between 90° and 180°, $\cos I = -\cos(180° - I)$.
„ I „ 180° „ 270°, $\cos I = -\cos(I - 180°)$.
„ I „ 270° „ 360°, $\cos I = +\cos(360° - I)$.

TABLE VI (cont.).

Giving value of $H \cos I$, where H is the semi-amplitude of the component and I the difference, in component time, between the time of high water of the partial wave and the time at which its height is required.

I \\ H	1·00	2·00	3·00	4·00	5·00	6·00	7·00	8·00	9·00	10·00
°	ft.	ft.	ft.	ft.	ft.	ft.	ft.	ft.	ft.	ft.
46	0·70	1·39	2·09	2·78	3·48	4·17	4·87	5·56	6·26	6·95
47	0·68	1·36	2·05	2·73	3·41	4·09	4·77	5·46	6·14	6·82
48	0·67	1·34	2·01	2·68	3·35	4·01	4·68	5·35	6·02	6·69
49	0·66	1·31	1·97	2·62	3·28	3·94	4·59	5·25	5·90	6·56
50	0·64	1·29	1·93	2·57	3·21	3·76	4·50	5·14	5·79	6·43
51	0·63	1·26	1·89	2·52	3·15	3·77	4·40	5·03	5·67	6·29
52	0·62	1·23	1·85	2·46	3·08	3·70	4·31	4·93	5·54	6·16
53	0·60	1·20	1·81	2·41	3·01	3·61	4·21	4·82	5·42	6·02
54	0·59	1·18	1·76	2·35	2·94	3·53	4·12	4·70	5·29	5·88
55	0·57	1·15	1·72	2·29	2·87	3·44	4·02	4·59	5·16	5·74
56	0·56	1·12	1·68	2·24	2·80	3·35	3·91	4·48	5·03	5·59
57	0·55	1·09	1·64	2·18	2·73	3·27	3·82	4·36	4·91	5·45
58	0·53	1·06	1·59	2·12	2·65	3·18	3·71	4·24	4·77	5·30
59	0·52	1·03	1·55	2·06	2·58	3·09	3·61	4·12	4·64	5·15
60	0·50	1·00	1·50	2·00	2·50	3·00	3·50	4·00	4·50	5·00
61	0·48	0·97	1·46	1·94	2·43	2·91	3·40	3·88	4·36	4·85
62	0·47	0·94	1·41	1·88	2·35	2·81	3·28	3·75	4·22	4·69
63	0·45	0·91	1·36	1·82	2·27	2·72	3·18	3·63	4·09	4·54
64	0·44	0·88	1·31	1·75	2·19	2·63	3·07	3·50	3·94	4·38
65	0·42	0·85	1·27	1·69	2·11	2·54	2·96	3·38	3·80	4·23
66	0·41	0·81	1·22	1·63	2·04	2·44	2·85	3·26	3·66	4·07
67	0·39	0·78	1·17	1·56	1·96	2·35	2·74	3·13	3·52	3·91
68·	0·38	0·75	1·13	1·50	1·88	2·25	2·63	3·00	3·38	3·75
69	0·36	0·72	1·07	1·43	1·79	2·15	2·51	2·86	3·22	3·58
70	0·34	0·68	1·03	1·37	1·71	2·05	2·39	2·74	3·08	3·42
71	0·33	0·65	0·98	1·30	1·63	1·96	2·28	2·61	2·93	3·26
72	0·31	0·62	0·93	1·24	1·55	1·85	2·16	2·47	2·78	3·09
73	0·29	0·58	0·88	1·17	1·46	1·75	2·04	2·34	2·63	2·92
74	0·28	0·55	0·83	1·10	1·38	1·66	1·93	2·21	2·48	2·76
75	0·26	0·52	0·78	1·04	1·29	1·55	1·81	2·07	2·33	2·59
76	0·24	0·48	0·73	0·97	1·21	1·45	1·69	1·94	2·18	2·42
77	0·23	0·45	0·68	0·90	1·13	1·35	1·58	1·80	2·03	2·25
78	0·21	0·42	0·62	0·83	1·04	1·25	1·46	1·66	1·87	2·08
79	0·19	0·38	0·57	0·76	0·95	1·15	1·34	1·53	1·72	1·91
80	0·17	0·35	0·52	0·69	0·87	1·04	1·22	1·39	1·56	1·74
81	0·16	0·31	0·47	0·62	0·78	0·94	1·09	1·25	1·41	1·56
82	0·14	0·28	0·42	0·56	0·70	0·83	0·97	1·11	1·25	1·39
83	0·12	0·24	0·37	0·49	0·61	0·73	0·85	0·98	1·10	1·22
84	0·11	0·21	0·32	0·42	0·53	0·63	0·74	0·84	0·95	1·05
85	0·09	0·17	0·26	0·35	0·44	0·52	0·61	0·70	0·78	0·87
86	0·07	0·14	0·21	0·28	0·35	0·42	0·49	0·56	0·63	0·70
87	0·05	0·10	0·16	0·21	0·26	0·31	0·36	0·42	0·47	0·52
88	0·04	0·07	0·11	0·14	0·18	0·21	0·25	0·28	0·32	0·35
89	0·02	0·03	0·05	0·07	0·09	0·10	0·12	0·14	0·15	0·17
90	0·00	0·00	0·00	0·00	0·00	0·00	0·00	0·00	0·00	0·00

Where I is between 90° and 180°, $\cos I = -\cos(180° - I)$.
„ I „ 180° „ 270°, $\cos I = -\cos(I - 180°)$.
„ I „ 270° „ 360°, $\cos I = +\cos(360° - I)$.

TABLE VII.

Harmonic Constants.

Place	M_2		S_2		N_2		K_1		O_1		K_2		P_1		M_4		MS_4		A_0
	H	k	H	k	H	k	H	k	H	k	H	k	H	k	H	k	H	k	
	ft.	°	ft.	°	ft.	°	ft.	°	ft.	°	ft.	°	ft.	°	ft.	°	ft.	°	ft.
South America, E. coast: Puerto San Antonio	10·31	302	2·15	17	2·16	260	0·65	302	0·34	227	0·55	19	0·19	300	0·17	268	0·04	285	13·44
Sea of Okhotsk: Langr island	0·92	181	0·20	222	0·23	96	1·31	275	1·15	246	0·07	222	0·46	275	—	—	—	—	3·78
China: Shanghai (Wusung)	3·11	30	1·03	77	0·40	2	0·66	207	0·46	149	0·28	77	0·22	207	0·70	331	0·47	18	6·55
Pakhoi	1·25	180	0·39	209	—	—	3·29	76	3·71	49	0·10	209	1·12	76	—	—	—	—	9·74
French Indo-China: Do Son	0·13	113	0·10	140	—	—	2·36	91	2·30	35	0·03	140	0·79	91	—	—	—	—	6·20
Philippine islands: Manila	0·66	305	0·22	338	0·13	291	0·97	320	0·93	279	0·07	325	0·30	317	0·02	352	—	—	1·40
Eastern Archipelago: Surabaya	1·44	351	0·85	355	0·30	337	1·54	318	0·89	284	0·26	357	0·46	321	—	—	—	—	—
Borneo: Labuan	0·89	321	0·39	359	0·20	299	1·35	315	1·08	266	—	—	0·46	322	—	—	—	—	—
Madagascar: Diego Suarez	2·10	111	0·95	155	—	—	0·39	55	0·26	63	0·26	155	0·13	55	—	—	—	—	4·36
Gulf of Aden: Jibuti	1·74	220	0·75	239	—	—	1·35	30	0·66	35	0·20	239	0·43	30	—	—	—	—	5·84
Africa, W. coast: Casablanca	3·35	47	1·25	77	—	—	0·26	36	0·20	293	0·33	77	0·10	36	—	—	—	—	6·30

TABLE VIII.

Giving the value of $\dfrac{(\text{Spring velocity} - \text{Neap velocity}) \times \text{Days from Spring tide}}{7}$,

Expression (125), according to the difference between velocities at springs and neaps and days from spring tide.

Days from Springs	Difference between velocities at springs and neaps															Days from Springs
	0·2	0·4	0·6	0·8	1·0	1·2	1·4	1·6	1·8	2·0	2·2	2·4	2·6	2·8	3·0	
0	0·0	0·0	0·0	0·0	0·0	0·0	0·0	0·0	0·0	0·0	0·0	0·0	0·0	0·0	0·0	0
1	0·0	0·1	0·1	0·1	0·1	0·2	0·2	0·2	0·3	0·3	0·3	0·3	0·4	0·4	0·4	1
2	0·1	0·1	0·2	0·2	0·3	0·3	0·4	0·5	0·5	0·6	0·6	0·7	0·7	0·8	0·9	2
3	0·1	0·2	0·3	0·3	0·4	0·5	0·6	0·7	0·8	0·9	0·9	1·0	1·1	1·2	1·3	3
4	0·1	0·2	0·3	0·5	0·6	0·7	0·8	0·9	1·0	1·1	1·3	1·4	1·5	1·6	1·7	4
5	0·1	0·3	0·4	0·6	0·7	0·9	1·0	1·1	1·3	1·4	1·6	1·7	1·9	2·0	2·1	5
6	0·2	0·3	0·5	0·7	0·9	1·0	1·2	1·4	1·5	1·7	1·9	2·1	2·2	2·4	2·6	6
7	0·2	0·4	0·6	0·8	1·0	1·2	1·4	1·6	1·8	2·0	2·2	2·4	2·6	2·8	3·0	7

TABLE IX.

Giving the value of $\sin \dfrac{h \times 180^\circ}{d}$, Expression (126).

Duration of stream (d)	Interval from commencement of stream (h)																					Duration of stream (d)
	0	½	1	1½	2	2½	3	3½	4	4½	5	5½	6	6½	7	7½	8	8½	9	9½	10	
3	0·0	0·5	0·9	1·0	0·9	0·5	0·0	—	—	—	—	—	—	—	—	—	—	—	—	—	—	3
3½	0·0	0·4	0·8	1·0	1·0	0·8	0·4	0·0	—	—	—	—	—	—	—	—	—	—	—	—	—	3½
4	0·0	0·4	0·7	0·9	1·0	0·9	0·7	0·4	0·0	—	—	—	—	—	—	—	—	—	—	=	—	4
4½	0·0	0·3	0·6	0·9	1·0	1·0	0·9	0·6	0·3	0·0	—	—	—	—	—	—	—	—	—	—	—	4½
5	0·0	0·3	0·6	0·8	1·0	1·0	1·0	0·8	0·6	0·3	0·0	—	—	—	—	—	—	—	—	—	—	5
5½	0·0	0·3	0·5	0·7	0·9	1·0	1·0	0·9	0·7	0·5	0·3	0·0	—	—	—	—	—	—	—	—	—	5½
6	0·0	0·3	0·5	0·7	0·9	1·0	1·0	1·0	0·9	0·7	0·5	0·3	0·0	—	—	—	—	—	—	—	—	6
6½	0·0	0·2	0·5	0·7	0·8	0·9	1·0	1·0	0·9	0·8	0·7	0·5	0·2	0·0	—	—	—	—	—	—	—	6½
7	0·0	0·2	0·4	0·6	0·8	0·9	1·0	1·0	1·0	0·9	0·8	0·6	0·4	0·2	0·0	—	—	—	—	—	—	7
7½	0·0	0·2	0·4	0·6	0·7	0·9	1·0	1·0	1·0	1·0	0·9	0·7	0·6	0·4	0·2	0·0	—	—	—	—	—	7½
8	0·0	0·2	0·4	0·6	0·7	0·8	0·9	1·0	1·0	1·0	0·9	0·8	0·7	0·6	0·4	0·2	0·0	—	—	—	—	8
8½	0·0	0·2	0·4	0·5	0·7	0·8	0·9	1·0	1·0	1·0	1·0	0·9	0·8	0·7	0·5	0·4	0·2	0·0	—	—	—	8½
9	0·0	0·2	0·3	0·5	0·6	0·8	0·9	0·9	1·0	1·0	1·0	0·9	0·9	0·8	0·6	0·5	0·3	0·2	0·0	—	—	9
9½	0·0	0·2	0·3	0·5	0·6	0·7	0·8	0·9	1·0	1·0	1·0	1·0	0·9	0·9	0·7	0·6	0·5	0·3	0·2	0·0	—	9½
10	0·0	0·2	0·3	0·5	0·6	0·7	0·8	0·9	1·0	1·0	1·0	1·0	1·0	0·9	0·8	0·7	0·6	0·5	0·3	0·2	0·0	10

Lightning Source UK Ltd.
Milton Keynes UK
UKOW03f0631041216

289103UK00002B/420/P